How to Use (and misuse) Statistics

Gregory A. Kimble is chairman of the Psychology Department at Duke University. He has published widely in the professional psychological literature and has contributed to the *Encyclopedia Americana* and the *Encyclopaedia Britannica*.

GREGORY A. KIMBLE

How to Use (and misuse) Statistics

PRENTICE-HALL, INC., *Englewood Cliffs, N.J. 07632*

Library of Congress Cataloging in Publication Data

KIMBLE, GREGORY A.
How to use (and misuse) statistics.

(A Spectrum Book)
Includes bibliographical references and index.
1. Statistics. I. Title.
HA29.K4875 519 5 77-27101
ISBN 0-13-436204-7
ISBN 0-13-436196-2 pbk.

A SPECTRUM BOOK

Printed in the United States of America

10 9 8 7 6 5 4 3

PRENTICE-HALL INTERNATIONAL, INC., *London*
PRENTICE-HALL OF AUSTRALIA PTY. LIMITED, *Sydney*
PRENTICE-HALL OF CANADA, LTD., *Toronto*
PRENTICE-HALL OF INDIA PRIVATE LIMITED, *New Delhi*
PRENTICE-HALL OF JAPAN, INC., *Tokyo*
PRENTICE-HALL OF SOUTHEAST ASIA PTE. LTD., *Singapore*
WHITEHALL BOOKS LIMITED, *Wellington, New Zealand*

To LLK and HP-65

Contents

Preface

When I was an undergraduate psychology major at Carleton College, longer ago than most of you are old, I took the path of least resistance through the curriculum. As there always will be, there were loopholes in the requirements and I managed to locate most of them. This meant that I graduated almost untouched by hard science, mathematics, and statistics.

Later I was to suffer for my sloth. As a graduate student, they made me take remedial courses to repair my deficiencies. And that experience dramatically changed my outlook on quantitative materials. Taking those courses, particularly statistics, working with the subject matter, and eventually teaching a course in advanced statistics on my first job at Brown University showed me the importance of what I had rejected as an undergraduate. My opinion now is that a course in statistics can be the most liberalizing course a student can find in the curriculum and perhaps should be required of everyone. The world we live in is a world of uncertainty. An acquaintance with the way a statistician thinks is as useful as anything I know in the struggle that we all must carry on in order to cope with what we cannot predict.

This last statement may provide a hint that I will not be much concerned with calculations in this book. What I want you to come away with is an appreciation of a style of thought and a respectable level of statistical literacy. I see no necessity, with these as my objectives, to dwell on formulas and computations. For those of you who find security in arithmetic, a final section of the book presents some of the technical tools. But what I want you to take away from your reading does not require mastery of that section.

I have tried very hard to keep the materials in this book lively and interesting, but I hope that I have not been *too* successful. By that I mean that it would be unfortunate if my routine flipness and irreverence and my occasional brushes with obscenity were to distract you and make you miss the serious points I have to offer. For I am really very serious about these materials. Understood, they will enrich your lives just as they have mine. So please: Read for understanding. If sometimes understanding is also fun, so much the better.

GREGORY A. KIMBLE

I wish to express my thanks to everyone
who helped me with this book, especially Norma Karlin,
Maxine Kluck, Tamara Easton, and Dan Thomann.

The Nature of Statistics

What is *statistics*? Ask an ordinary person and you are apt to get a response that reflects Disraeli's opinion that "there are three kinds of lies: lies, damned lies and statistics." Ask a statistician and the answer will be different: Statistics is a branch of science dealing with the collection, analysis, and interpretation of data. As the title of this book tells you, both answers suggest topics worth writing about. Although the popular notion that you can "prove anything with statistics" is far too extreme, it is possible to misuse statistics in order to mislead people. But statistics can also aid your understanding of many of the problems that face the world today. The pages to follow will present examples of both of these uses of statistics. It is useful to think of statistics as serving two general functions, a descriptive one and an interpretive one. This chapter presents an overview of both functions.

DESCRIPTIVE STATISTICS

Descriptive statistics are exactly what their name implies, numbers that describe some situation of interest. Batting averages, rates of unemployment, rates of mental illness, sales of automobiles in November, number of children in the average family, average income, frequencies of sexual activity, and the well-known tables of heights, weights, and lengths of life are all examples.

The value of descriptive statistics is that they give an efficient summary of some type of information. A professional baseball player may come to bat 500 or even 600 times during a season. A list of the player's performance on every such occasion would be so long and complicated as to make it impossible to obtain a very clear overall impression of the record. In 1976 Carew of the Minnesota Twins came to bat 605 times and got 200 hits. A complete record of all Carew's at-bats, together with an indication of what happened on each occasion, would take pages. The important information can be conveyed briefly, however, by dividing 200 by 605 and reporting that Carew's batting average was .331.

Batting averages are straightforward and uncomplicated because the operations involved in their computation are completely clear. We know what it means to be at bat and to get a hit. We also know what it means to divide one of these terms by the other. Sometimes, however, things are not so clear.

Sizes of Cities and How They Grow

When you look up a city's population in *The World Almanac*, you will find a comfortably definite number. For Boulder, Colorado, where I live, the number is 66,870. But what exactly does this mean? For example, does it include students at the university? Probably not, unless they have established residence in Boulder. How many additional people does this add to the population? There is no real way to answer the question because a student at the university turns out to be a statistical abstraction. The university operates under a legislated limit of 20,000 students per year. But this does not mean 20,000 human beings; it means 300,000 student credit hours. For purposes of limiting the size of the university, a "student" is 15 credit hours of work. Since the average student takes fewer than 15 credit hours of courses, the number of people is some indefinite number greater than 20,000. Some unknown fraction of them (those who are not Boulder residents) should be added to 66,870 to produce a more accurate figure for the population of the city.

The estimate of Boulder's population is complicated further by the fact that whether a person lives in Boulder is an arbitrary matter, because city limits are set arbitrarily. In the case of our city this fact is important, because a very large number of people who are Boulder residents for most purposes live beyond the city limits. Various ways of estimating the number of such people suggest that there are 20,000–25,000 of them. These figures, together with a guess that perhaps 15,000 students should be added to the population, indicate that the true population of Boulder is probably closer to 100,000 than to the almanac figure of about 67,000.

Similar ambiguities exist in reports of the rates of growth for cities. To make this point, I shall use some statistics for Raleigh and Durham, North Carolina, where I lived before I moved to Colorado. Populations of the two cities for 1960 and 1970, along with some statistics that can be derived from them, appear in Table 1–1.

Obviously, Raleigh is bigger than Durham. It also appears that Raleigh is growing faster. But again there is more to the situation than meets the eye. Both cities are university cities and exact population figures depend upon how students are counted. Once more there is also a problem relating to the setting of city limits, which accounts for some of the apparent difference in rates of growth. If my memory is correct, Raleigh extended its

TABLE 1–1
Populations of Raleigh and Durham, North Carolina, 1960 and 1970

City	1960 Population	1970 Population	Increase in Population	Percentage Increase in Population
Durham	78,302	95,438	17,136	22
Raleigh	93,931	123,793	29,862	32

limits sometime in the 1960s but Durham did not. If Raleigh's expansion brought an additional 9,218 people into the city, this would account for the difference between a 22% and a 32% increase in the sizes of the two communities.

It may be important to point out that all of this is not just an idle exercise designed to get across statistical ideas. Such information has great practical significance. A merchant considering Boulder as a place to open a branch store might make quite different decisions depending upon whether he believes that Boulder's population is 67,000 or 100,000. Recognizing such concerns, the U.S. Office of Management and Budget (OMB) has defined Standard Metropolitan Statistical Areas (SMSAs) on the basis of census data and reports populations for 266 urban areas. From time to time it also combines areas when they seem to represent a single economic unit. The results can be spectacular. In 1970 the Raleigh SMSA had a population of 228,453 and a rank in the nation of 135. By 1975 the OMB had combined regions and created a Raleigh/Durham SMSA. It has a population of 462,300 and is 78th in the nation.

Unemployment

For a good many years, the following question has been high in the minds of everyone: "Are we in a recession or not?" Whether we are or not depends upon the definition of a recession. Several economic indices, including production in heavy industry, the sizes of inventories in retail stores, and rate of unemployment provide possible bases for such a definition. The definition in terms of unemployment seems particularly relevant to human welfare. As a result, most of us pay some attention to the statistics on unemployment and take these figures as an important index of the economic health of the nation, possibly saying to ourselves that if the unemployment rate exceeds 9%, the nation's economy is in a bad way—we are in a recession. To take this position is to define a time of recession as a time when more than 9% of the population is out of work.

Thus the answer to the question, "Are we in a recession?" depends upon the answer to the more specific question, "Is the rate of unemployment more than 9%?"

In November 1975 official U.S. government figures indicated that the jobless rate was 8.3%. Albert Sindlinger, a market research analyst, criticized this figure, claiming that it should have been over 9.2%, perhaps as high as 10.6%. In short, by the definition of a recession as 9% unemployment, we were in one according to Sindlinger but not according to the official figures. How could such a disagreement come about?

To answer this question, it is important to understand how the percentage of unemployment is determined. The calculation is almost the same as the computation of a batting average. It is the number of people without jobs divided by the total number of people in the labor force. In November 1975, the official computation was: 7,717,257 unemployed divided by 92,979,000 in the labor force equals .083, or 8.3%.

The difference between this calculation and that of a batting average is that both of the basic terms in the calculation are a little vague. Of the two, the concept of labor force is the more definite, and Sindlinger agreed with the official figure in the dispute mentioned above. The labor force is everyone over 16 who might be expected to work. During the school year it does not include students unless they hold a job, but it does include them during summer vacations. Thus the size of the labor force increases by a few million each June and a "seasonal adjustment" enters the calculation of the rate of unemployment. A second major group omitted from the labor force are housewives. They enter the force only if they start looking for a job. Whatever one makes of this treatment of certain groups, the definition of labor force is precise enough to create no great problem.

The definition of an unemployed person is much more difficult. One definition might be everyone who is looking for a job. This definition would then include everyone who has a job but is trying to find a better one. Defined that way, the unemployment rate may typically be as high as 11 or 12%. Or one could go to the other extreme and say that a person is not really unemployed unless he has been out of work for a while—15 weeks being a period for which 1975 data are available. Defined in those terms, the percentage of unemployed was only about 3.3%.

Obviously, the jobless rate is a more elusive figure than one might have thought. Looking further into the definition of unemployment only complicates matters. Does part-time work count as employment? If so, is an hour's work a week on a college work–study program enough to qualify a person as employed, or must it be more? How about people on the Public Service Employment Program who have jobs created by the government specifically for the unemployed? Should they be counted or not? The list could go on. I mention these two items because they appear to have been responsible for the dispute between Sindlinger and the Feds. If the official count of the number of employed people included 700,000 work–study students and 315,000 Public Service Employees (reasonable estimates) in November, and if they should not have been counted as employed, the rate of unemployment was more like 9.4% than the reported 8.3%. It is the elusiveness of such figures that allows them so frequently to come up for argument in politics, for example in the Carter–Ford debates of 1976.

Rates of Mental Illness

Those who believe that the world is going to hell in a hand-basket will tell you that this unhappy trend includes an increase in the rate of mental illness. The greater complexity of life and the resulting greater stress are to blame, according to this view. Deciding the truth of this assertion poses very special problems. For one thing, in order to detect the alleged change it would be desirable to make a comparison between rates of mental illness for widely separated points in time—ideally 75 or 100 years. Such a comparison might, for example, look at admission rates (proportions of the population admitted to mental hospitals) as revealed by older hospital records and newer ones. Such studies have been done and superficially the data suggest that a trend toward increased frequency of mental illness actually exists.

There are several things wrong with these[1] data, however:

1. Mental illness is diagnosed more frequently now than it was 100 years ago. Complaints now seen as mental disorders once were not. This means that the comparison must be for such disturbances as schizophrenia, which have been identified by the same symptoms for about as long as psychiatric diagnosis has been around.
2. Whatever the disorder, the increased availability of mental hospitals means that more people are undergoing treatment now.
3. The population now contains a greater proportion of old people and they have a higher incidence of mental disorder.

Correction for these features of the data reveals that the incidence of mental disorder is probably no different now from what it was in the middle of the last century.[2]

[1]Along with Edwin B. Newman and others, I am fighting what appears at the moment to be a losing battle against a variety of grammatical atrocities. *Data* is a plural word. It refers to more than one item of information. The singular is *datum*. Thus you should speak of "these data" and "this datum."

[2]W. A. Wallis and H. V. Roberts, *The Nature of Statistics* (New York: Free Press, 1962).

Crime Rates

Increases and decreases in crime rates are also harder to prove than you might think, for several reasons:

1. One basis for determining the number of crimes is the number reported, but if people change in their willingness to report crimes, a change in rate could be detected where none existed.
2. Another basis for defining crime rate is the number of individuals caught and booked for criminal acts. If the police forces improve and more criminals are apprehended, the crime rate may erroneously appear to increase on this basis.
3. In a similar way, better systems of keeping criminal records will produce an increased crime rate that is apparent but not real.
4. The definition of "crime" will influence the rate at which crimes occur. In some times and in some places, homosexuality and the possession of small amounts of marijuana are major crimes, felonies. If the laws change and these infractions become misdemeanors rather than felonies, the felony rate will decline.

OPERATIONISM AND SOME RELATED IDEAS

If you are willing to put up with a brief discussion that is slightly philosophical, the materials just presented can be made to illustrate some very important points about straight thinking on many topics. My plan for this section is to develop the essential ideas in a position called *operationism*. I hope to convince you that the straight thinking just referred to is operational thinking. In these antiintellectual times, such thinking has been subjected to a great deal of bad mouthing. If I am successful in my effort, you will come away from your reading of this section inoculated against the effects of such mindless criticism.

Operationism is a general view of science which holds that scientific statements are meaningful only if they lead to observations that can be made on the real world. Otherwise, these statements are meaningless. A key concept in the operationistic position is that of operational definitions, which are definitions of concepts in terms of physical procedures.

In the most general sense a definition is a statement that gives the meaning of a word or group of words. The last sentence is an example. It is a definition of the word *definition*. Definitions of this type pose a small problem. They make no sense if you happen not to understand the terms employed, for example "gives the meaning of." To make this point more realistically, suppose I tell you (correctly) that "a *Z* score is the deviation of a raw score from the mean expressed in units of standard deviation." This definition is meaningless to almost everyone, because most people do not know the meaning of such terms as *raw score*, *mean*, and *standard deviation*.

Now contrast this with either of the definitions of an unemployed person as an individual (1) who says he is looking for work, or (2) who has been without a job for three months. Obviously, these definitions are less complex than the definition of Z score, but for our purposes that is not the important difference. The important difference is that the definitions of unemployed person tell you how to identify such an individual. They identify the procedures or operations that you would use. For this reason such definitions are called *operational definitions*.

Most of the materials presented in the first section of this chapter were little exercises in operationism. In each example the basic concern, you will now see, was about the operational definition of some concept. To review:

1. The population of a community depends upon who is defined as living there.
2. Growths in population depend upon sometimes arbitrary definitions of the area for which the population is counted.
3. The rate of unemployment depends upon the definitions of the (a) labor force and (b) an unemployed person.
4. Calculated changes in rate of mental illness depend upon record-keeping operations.
5. The same is true for crime rates.

Operationism and Common Sense

The general point to get out of all this is that exactly what a statement of alleged "fact" means depends upon the operations carried out to obtain that "fact." Sometimes it turns out that such operations are impossible or nonexistent. A couple of examples will illustrate this state of affairs.

Settling Arguments

Obviously, it is very easy for arguments to arise regarding the sizes of cities and what is happening to the crime rate, to the rate of mental illness, or to the rate of unemployment. In the last case, one person might argue that the rate of unemployment is 10 or 12%; another might claim that it is only 3 or 4%. We have already looked at the possible bases for the various claims and have seen that it is all a question of who is counted as jobless. If an unemployed person is anyone who is looking for a job, the rate is high; if it is anyone who has been out of work for three months, the rate is low. In other words, it all depends on the *definition* of unemployment. If the two individuals quoting different numbers for the rate of unemployment will accept the same definition of unemployment, the issue can be settled. If they will not accept the same definition, at least they will see that the argument is about definitions rather than about the facts of the situation.

Silly Statements

My students in Introductory psychology these days tell me with some frequency that it is well known that people never use all of their "brain power" but actually use only 10% (or some other small but exact percent) of it. They usually want to know why this is so, hoping to increase their own mental effectiveness. Unfortunately, the assertion is operational nonsense:

1. How do you define "brain power?"
2. What kind of measurement is involved that will permit quantitative statements?
3. If 100% of it is never used, how could you possibly know that 10% is all that is used? (To know the value of a fraction demands that you know the value for the whole.)
4. If people do function below their best levels (perhaps true), why blame it on brain power rather than on poor motivation, sloppy work habits, or something else?

The new personalistic, dynamic, humanistic discussions of mental life are full of similar operationally silly statements. To take just the most recent example to come to my attention, Dorothy Tennov has this to say in her marvelous book, *Psychotherapy: The Hazardous Cure*.[3] "The phrase 'get in touch with your feelings' drones throughout the phenomenological litany with maddening insistence. One speaker used it fourteen times in as many minutes. It is a curious phrase because it flouts scientific examination. The phrase implies that one has feelings one does not feel!"

Finally, let us take one small example from the advertisers whose announcements will reward the attention of an operational eagle eye. The TV promotion of one of the major fast-food chains tells us that every one of their burgers contains 100% pure ground beef. So it almost certainly does, but there is nothing in the statement to guarantee that there is more than one-tenth of 1% of it.

[3]D. Tennov, *Psychotherapy: The Hazardous Cure* (Garden City, N.Y.: Doubleday, 1976).

INFERENTIAL STATISTICS

The foregoing discussion has strayed pretty far from descriptive statistics. I will come back to the topic in later chapters. Now, however, I would like to introduce you to the other half of the statistical enterprise, *statistical inference*, the making of wise decisions in the face of uncertainty.

Benjamin Franklin is supposed to be responsible for the observation that nothing is certain in this life but death and taxes. I would like to add a third certainty to the list, the certainty of uncertainty. We live in a probabilistic world, and most of the decisions we make entail an element of risk. If I am a merchant, what proportion of dresses with long and short skirts shall I stock? Some women will wear long skirts, some short, but in what ratio? If I am a gambler, shall I draw to the straight, keep a kicker, bet on the favorite, or finesse the queen? If I am a student taking a multiple-choice test, shall I pick choice "b" or "d"? I don't really know the answer but I suspect that one of them is right. If I am a sick man, shall I submit to surgery? The doctor says that it will probably cure me but there is some chance that I will not survive the operation.

Obviously, there is more to consider in such cases than just the odds, and certainly statistics cannot solve important human problems. But in many cases statistics can provide considerable clarification and a basis for decision. That is what inferential statistics is all about. In this section, I shall develop the essential ideas in this area, beginning with a frivolous example.

The Riddle of the Neglected Lover

To the best of my knowledge, the following puzzle is the invention of Harvard statistician Frederick B. Mosteller.

The scene is a large city and the point of the story involves its subway system. This system is in the form of a continuous loop on which a single train travels, always in the same direction. In general layout the system resembles a figure eight or a dumbbell, *perhaps* (please note this) as shown in Figure 1–1. The main station in the system is at the point where the two sides of the loop are close together. At this point, by going to one side of the station or the other, a traveler can catch the train going in either direction. Other stations are located at intervals along the rest of the track, two of these other stations figure in the events to be described now.

Our concern is with three people who are served by this system: Mr. Z, whose office is near the main station, and his two lovers, Ms. A and Ms. B, who live on opposite sides of the city in different sections of the system.

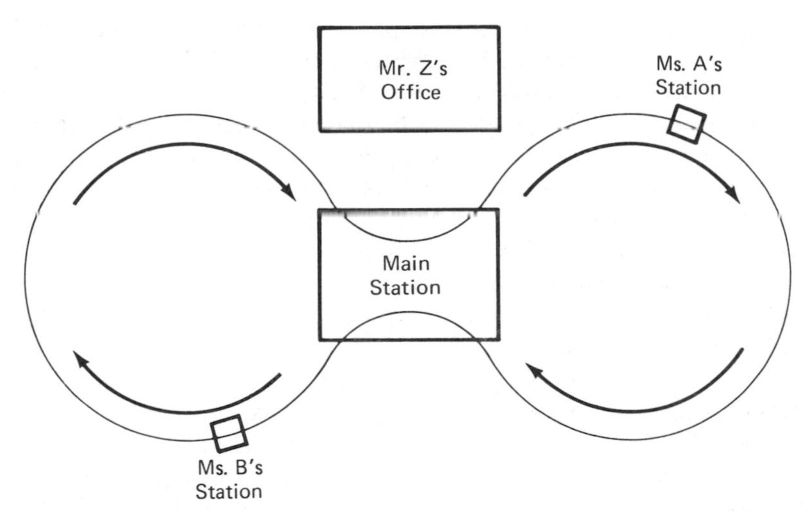

Figure 1–1 City and subway system imagined by Mr. Z. If this conception is correct and if Mr. Z comes to the main station at random times, he should be as likely to catch the train going one way as the other.

Mr. Z has a problem. Ms. A and Ms. B are equally attractive to him. He visits one or the other of the women on a daily basis, but the choice of which one to visit is always difficult. His solution to the problem is to "leave things up to chance." When he can get away from work (which happens at times that vary unpredictably from day to day), he goes to the station and boards the train whenever it comes. If the train is going east, he visits Ms. A; if it is going west, he visits Ms. B. On the assumption that the probability of catching the east-going train is about the same as that of catching a west-going train, the visits should even out in the long run.[4]

But things do not work out that way: after a period of several weeks Ms. B complains that she is being neglected. Somewhat in disbelief Mr. Z puts Ms. B's complaints to a test and keeps a record of his visits for a month. The data support Ms. B's allegations. There were 21 visits to Ms. A during the month but only 9 to Ms. B.

Was It the Fickle Finger of Fate? How could such an unfair outcome happen? Before offering a possible explanation for the unjust way in which chance has dealt with Ms. B it will be important to look at the puzzle a little more closely. Although this fact cannot be clear to you yet, by the time you finish this book you will understand that Mr. Z's dilemma

[4]I hope that you will not be put off by the male chauvinistic, heterosexual character of this example. If it offends you, the genders of the players in my little drama can be altered to suit your preferences. The logic of the argument will still hold and that is what I want you to get from this section. Another way of thinking would have Ms. A making the complaint. Again, however, the statistical reasoning I am about to describe will work.

presents the basic structure of the problems handled by the standard methods of testing statistical hypotheses. The only difference is that the quantitative aspects are minimal in this example.

Consider first Mr. Z's reasoning when he decided to leave the choice of visits to chance. In the long run, he argued, the number of visits to Ms. A and Ms. B should be the same because of the equal probability of catching a train going in either direction. There should be *no difference* between the number of visits. This hypothesis of "no difference" technically goes by the name of the *null hypothesis*. For the specific test that Mr. Z decided to carry out, the null hypothesis would be that he would make 15 visits to Ms. A and 15 visits to Ms. B.

Exactly 15? Well probably not. Our intuitive sense of the "law of averages" leads us to expect something other than an exact 50–50 split. How about the 21–9 split that actually occurred? Probably most readers will be inclined to reject the idea that such an unequal division would happen under circumstances where the expected split is 15–15. To put the point more technically, such readers *reject the null hypothesis* and conclude that something must have been wrong with Mr. Z's assumptions.

To develop this point just a bit further, suppose that the visits had been split 17–13. What would you make of that? Probably that such an outcome is within the bounds of chance. How about 18–12 or 19–11 or 20–10? Somewhere you find yourself saying, "Too much already! I doubt that this could happen by chance. I reject such an hypothesis. There must have been some bias in favor of taking the A train."

Making Wise Decisions in the Face of Uncertainty. But when such seemingly unlikely outcomes force you to such a conclusion, is the conclusion the right one? This is a question that we cannot answer with a definite yes or no. What we can say is of fundamental importance, however.

The 21–9 split *could* have occurred by chance even if the true odds of catching Ms. A's train or Ms. B's train were exactly the same. If you were to toss 30 fair coins, it *could* happen that 21 of them would turn up heads and that only 9 of them would turn up tails. Such happenings are rare. It can be calculated[5] that 21 of 30 coins will come up heads (or tails) less than 5 times in 100. But such rare events do occur, and this is why an absolute yes–no decision is never possible in a case like Mr. Z's. There is always a choice between two interpretations:

[5]You will find as you get further into this book that I have not the slightest interest in explaining the computational procedures that are commonly used to bolster statistical arguments. I have avoided such explanations whenever I could on the general philosophy that there are other ways to get my ideas across. On the other hand, the reader has the right to be sure that I am not just pulling the numbers out of thin air. For that reason I have included a computational appendix (pp. 221–54) that presents formulas and makes many of the calculations mentioned in the text.

1. The situation *is not* what it was originally assumed to be (the null hypothesis is wrong) and the results obtained are a better reflection of the true state of affairs.
2. The situation *is* what it was originally assumed to be (the null hypothesis is right) and we are faced with one of the rare chance occurrences just discussed.

Types of Error. Faced with the alternatives just presented for his particular problem, Mr. Z could argue this way: If the chances of catching an east-bound or a west-bound train actually are 50–50 (if the null hypothesis is true), a split as extreme as 21–9 would happen by chance only 4 or 5 times in 100. Since such an outcome is so unlikely, it makes better sense to consider alternatives to the 50–50 null hypothesis (to reject the null hypothesis) in favor of other possibilities.

What could some of these possibilities be? In general, anything that keeps the train in Ms. B's part of town longer than in Ms. A's would increase the probability that when the train came it would be going in Ms. A's direction. Possibly the two segments of the route are unequal in length, as shown in Figure 1–2A. Possibly the segments are of the same length but there are more stations and thus more time-consuming stops in one section, as shown in Figure 1–2B. Possibly the track is inferior in Ms. B's segment and the train must go more slowly. Given these possibilities and the data at hand, Mr. Z would be in good statistical company if he were to decide that his 50–50 null hypothesis must be wrong and that it must be rejected. Pushed on the point Mr. Z could even reject the null hypothesis at a certain *level of confidence*. He could reason this way: "Well, I could be wrong, but the chances are only 4 or 5 in 100 that I am wrong. So I will reject the null hypothesis at what I will call the 4 or 5% level of confidence."

But, as he recognized himself, Mr. Z *could* be wrong. This *could* be one of those rare chance occurrences. If that were the case, Mr. Z would have made what we will call a *Type I error*. Although with good reason, he has rejected the null hypothesis when it is, in fact, true.

Now consider another possibility. Suppose that the true situation actually is one of those shown in Figure 1–2 but, for whatever reason, such possibilities do not occur to Mr. Z. Unable to think of the alternatives, he reasons that the null hypothesis is true (the probability of catching one train actually is the same as the probability of catching the other) and that his pattern of visits is just one of those rare chance happenings. In this case Mr. Z has made a *Type II error*: he has accepted the null hypothesis when it is false.

In common with many of us, Mr. Z might go on to reason that after such a long string of trains that took him to Ms. A, the "law of averages" will take care of things and there will be a succession of trains to Ms. B

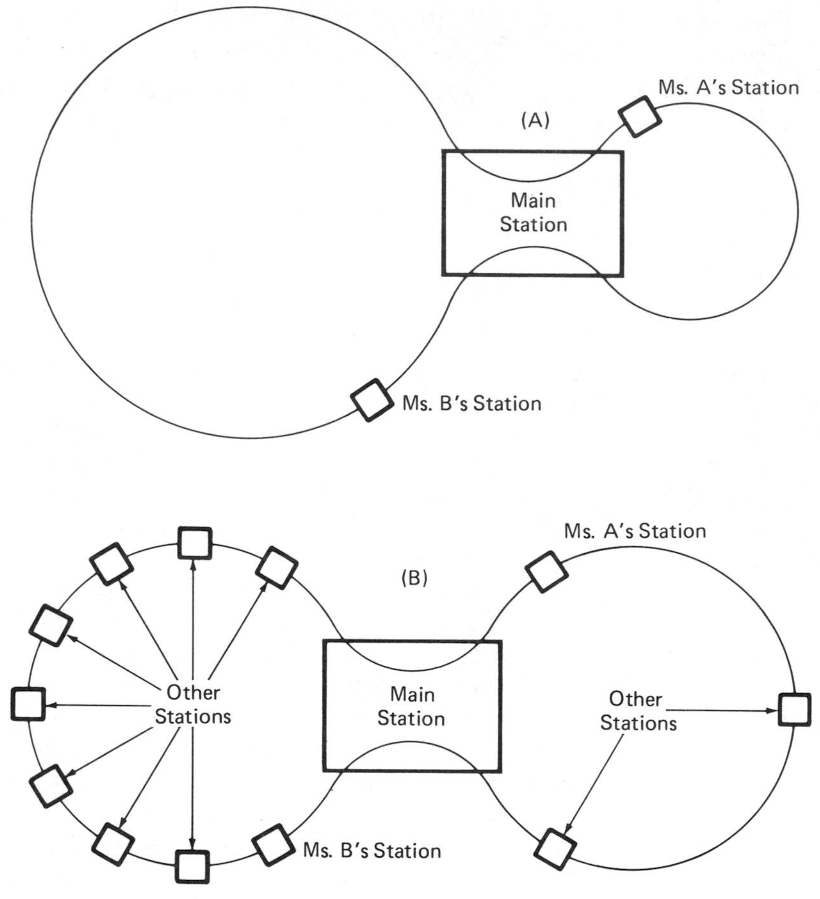

Figure 1–2 Alternatives to Mr. Z's conception of the city that might account for the results by keeping the subway train in Ms. B's section of the city for longer than it is in Ms. A's section. Thus increasing the probability that the train will be leaving Ms. B's section and going in the direction of Ms. A.

that would even the number of visits out. If he does take this further step in his thinking, Mr. Z has fallen prey to a very different kind of error: the "gambler's fallacy." More of that later (p. 100).

To Put It Briefly

Believe it or not, the ideas presented in the past few pages take you through about half of a first course in statistics. One thing I hope you see is that the ideas are not very difficult. Another thing I hope you see is that statistics is not so much about mathematics, formulas, and calculations as it is about ways of reasoning. In this section, I would like to summarize such reasoning as it applies to the making of inferences from data. The summary involves a series of steps that apply to a wide range of statistical tests. Now, with the aid of our example, the steps:

1. *State the hypothesis in null form*: Except for the accidents of chance, there is no difference between the number of visits to Ms. A and Ms. B. In 30 days they should divide 15–15.

2. *Obtain data of the kind identified in the null hypothesis*. Twenty-one visits to Ms. A, 9 to Ms. B.

3. *Determine the chance probability of occurrence (4 or 5 times in 100) of the data obtained if the null hypothesis is true*. This is the step in reasoning where the example left procedures pretty vague. I will explain them more fully beginning on page 126. The probability in question, however, is only 4 or 5 in 100.

4. *If the chance probability of the obtained result is small, reject the null hypothesis with a level of confidence that is the probability of obtaining the result by chance*: 4 or 5 times in 100. If the chance probability is high, accept the null hypothesis.

5. *Recognize that either rejecting or accepting the null hypothesis involves a gamble*. The null hypothesis allows the occurrence of the 21–9 split 4 or 5 times in 100. If the statistician rejects this null hypothesis because of long odds against an outcome in spite of the fact that unknown to him the null hypothesis is true, he has committed a Type I error. But if he accepts the null hypothesis when it is false, the statistician commits a Type II error.

Probably the most unfamiliar concepts in this section are those of Type I and Type II error, and it may be worth enlarging our review of them. One way to get the distinction across is by analogy to the decision of a jury in a criminal case. By our laws a defendant is innocent (null hypothesis, he did not do it, step 1 above) until proved to be guilty (by evidence or data, step 2) beyond a reasonable doubt (with a strong level of confidence, steps 3 and 4). Now think of the two possible mistakes that a jury can make. It can find an innocent person guilty (Type I error) or a guilty person innocent (Type II error). The other two decisions the jury can make are correct decisions and, therefore, just ones.

This interpretation leads to still another way to make the distinction. Table 1–2 presents the true status of the null hypothesis in its columns and the statistically arrived at decision in its rows. The cells produced by crossing rows with columns contain an indication of the significance of each of the four possible outcomes. If the table is unclear, one way to straighten things out would be to make a similar table for the decisions of juries, with the columns representing "truly innocent" and "truly guilty"

TABLE 1–2
Correct Decisions and Types of Error in Hypothesis Testing

Statistical Decision	Real-World Status of the Null Hypothesis	
Accept null hypothesis	True: Correct decision	False: Type II error
Reject null hypothesis	Type I error	Correct decision

and the rows representing jury decisions of "acquittal" and "conviction." The meaning of the four cells will then correspond to the meanings I have put in the table in this book.

Research Hypotheses

Looking ahead, it seems important to add one brief point before we leave this topic. Actual research usually begins with what is called a *research hypothesis*. The investigator believes that some condition he can control will have an effect on some phenomenon that he is interested in. He collects data hoping that the expected difference will occur. In order to obtain support for this research hypothesis, the investigator actually tests the null hypothesis. If the null hypothesis can be rejected, the research hypothesis gains in credibility.

PARAMETER ESTIMATION

In order to round out this preview of things to come, I would like to give you a brief glimpse at the process of *parameter estimation*. Be prepared, as usual, to find out that what goes on in parameter estimation is a lot less difficult than the expression leads you to expect. A parameter is the value of some measure for a population. A *population* is the entirety of some collection of things: all the people in some community, all the temperatures for August 14 in San Francisco since such records have been kept, or all the tea in China. Usually our interest is in measures on populations, that is, in *parameters*. Average income and variability of income in the working population of the nation would be examples. It is a rare thing for information to be available on an entire population. More often the only information available is on a *sample* that constitutes only a small fraction of the population. Such measures obtained on samples are called *statistics*. One usefulness of these statistics is that they allow us to infer the value of the parameter.

The Fishes of the Sea

A possibly important example of parameter estimation, with the world's food supply running low, could involve an estimate of the number of edible fish in all our oceans. I can illustrate the method with a more modest project of estimating the number of fish in a single pond. There is a standard way of doing it. Catch a sample of (say) 100 fish and mark them all with a metal tag. Catch another sample, let us say of 100 again, and note the number of tagged fish in the second sample. Suppose there were 10. This suggests that the probability of being caught is 10 in 100, or 1/10, and that there must be ten times as many fish in the pond as were in the sample of 100, that is, 1,000.

It is very important to note one thing about this procedure, the assumption that every fish in the pond stands an equal chance of being caught. A sample where every individual stands an equal chance of selection is a *random sample*. This "assumption of randomness" is basic to many statistical procedures.

Forecasting the Outcomes of Elections

The parameter estimated in the previous example was the size of a population. A different type of estimate involves the proportion of a population with a particular characteristic, for example, casting a vote for the Democratic candidate in the next presidential election. The general idea behind the making of such estimates is even simpler than that involved in estimating the size of a population. The opinion pollster goes to a representative sample of voters, finds out the fraction who will vote Democratic (statistic), and uses this as an estimate of the population value (parameter).

Errors of Estimate

The inference that there are 1,000 fish in a pond or that 54% of the voting population will vote Democratic in the next election are *point estimations*, because they involve a single number or point on some scale. It is easy to see that such estimates are not likely to be exactly right. A different sample of fish or voters would almost certainly lead to an estimate that is slightly different from the first one. For such reasons, estimates of parameters almost always come with a *confidence interval* attached. The pollster will say that the best estimate of the percentage of people who will vote Democratic is 54% and that one can be 95% confident that the percentage will be in the interval 52–56%, and 99.9% confident that it will be in the interval 49–59%. How confidence intervals are arrived at is something I will consider later.

SUMMARY–GLOSSARY

In addition to the occasional interim summaries that appear within chapters of this book, I will bring each chapter to a close with a final summary which, in large part, will be a glossary that reviews the important concepts from each chapter. In the case of this first chapter, the following are the terms that you must be sure you understand.

Statistics. A branch of science dealing with the collection, analysis, and interpretation of data.

Descriptive statistics. Measures that summarize the characteristics of a set of data.

Operationism. The view that scientific statements must be relatable to objective operations.

Operational definition. A definition that relates a concept to objective operations rather than simply to other words.

Inferential statistics. Procedures for estimating the characteristics of populations from data on samples.

Null hypothesis. Hypothesis of "no difference" set up for statistical test. In research the investigator usually hopes to be able to reject the null hypothesis, in this way obtaining support for some research hypothesis that gave rise to the research. Testing the null hypothesis is an example of inferential statistics. If the null hypothesis is rejected, the inference is that the population difference is not zero.

Level of confidence. The probability of obtaining a result as different from zero, as was obtained in an investigation if the null hypothesis is true.

Type I error. Rejecting the null hypothesis when it is true.

Type II error. Accepting the null hypothesis when it is false.

Research hypothesis. A factual hypothesis that is tested by empirical investigation.

Parameter estimation. The process of estimating population values from sample values, that is, of estimating parameters from statistics.

Population. All the members of an actual or theoretical collection of persons, objects, or items from which samples are drawn for statistical purposes.

Parameter. A measure for a population.

Sample. A subset of the members of a population.

Statistic. A measure obtained on a sample, usually for purposes of estimating the value for a population.

Random sample. A sample in which every individual and every combination of individuals stands an equal chance of being selected.

Point estimation. The exact value of a parameter estimated for a population.

Confidence limits. The range of values on either side of an estimated point that contains the true value of the parameter, with a stated degree of certainty.

Pictures of Data

A picture, they tell us, is worth a thousand words. Although putting the point that precisely is of dubious value and even the truth of the observation is sometimes questionable (for me, at least, Carl Sandburg's seven-word metaphor, "The fog comes on little cat feet" says more than any misty picture I can remember), it is true that pictorial representations can be eloquent. The concern of this chapter is with the pictorial representations of data, that is, with charts, graphs, and the like.

In my experience people seem to have two kinds of reactions to such materials. For some, graphs come naturally. For these individuals a graphical presentation is the clearest, most efficient way possible to get an idea across. For others, graphs are a big mystery. This chapter is addressed to members of the second group. I think that, with a little practice, understanding graphs can come to be sort of second nature to you, too. If you will learn a few simple rules about the way in which graphs are made, all the mystery should disappear. For readers who are at home with graphs, I will try to keep things interesting by the selection of examples. Even if you know all about these materials, there should be something of value in the illustrations I use to make my points.

PROLOGUE: EXPLANATION IN SCIENCE

A common way of thinking has it that the goal of science is to explain the phenomena of the natural world. Unfortunately, discussions of what it means to explain something come forth with various positive definitions of *explanation*. There is more agreement on one thing that explanation is not: it is not merely the naming of phenomena. Although this point is probably obvious, there are occasional problems in recognizing that an alleged explanation really is nothing but naming. Criticizing the concept of instinct in these terms, E. B. Holt once put the point in colorful language:

> Man is impelled to action, it is said, by his instincts. If he goes with his fellows, it is the "herd instinct" which activates him; if he walks alone, it is the "anti-social instinct"; if he fights, it is the instinct of "pugnacity"; if he

> defers to another, it is the instinct of "self-abasement"; if he twiddles his thumbs, it is his thumb-twiddling instinct; if he does not twiddle his thumbs, it is his thumb-not-twiddling instinct. Thus everything is explained with the facility of magic—word magic.
>
> One could have hoped that Molière, in the seventeenth century, had given the *coup de grâce* to such verbal tomfoolery, in that familiar passage where the candidate in medicine says: "I am asked by the learned doctor for the cause and the reason why opium induces sleep. To which I reply, because there is in it a soporific virtue whose nature it is to lull the senses."[1]

Another way to describe this criticism is to say that the explanation is *circular* or that the thing to be explained and the "explanation" are the same thing. In the opium example the fact to be explained is that opium puts people to sleep. "Soporific" (the most important term in the explanation) means "sleep inducing." Thus opium puts people to sleep because it is sleep inducing. That is, it puts people to sleep because it puts them to sleep.

Even 300 years later, medical terminology and sometimes the medical profession need an occasional slap on the wrist for substitution naming for explanation. The latest abuse involves the diagnosis of *dyslexia* (which may be made more often by school psychologists than by physicians). In any event, we heard first about the fact that Johnny can't read and then that Johnny can't read *because* he is dyslexic. Quasi-medical verbiage! No better than the opium example. *Dyslexia means trouble with reading*, and to offer this as an explanation is to mislead. Moreover, if the parents of a "dyslexic" child look further into the matter, they may run across an earlier definition of the term which limited dyslexia to reading problems resulting from a very rare kind of brain damage.[2] The consequences for the child and the parents' conclusions about their responsibility or lack of it are frightening to contemplate.

Some other "explanations" of the same nonexplanatory type are the following. Johnny won't obey *because* he is in a negativistic stage. Baby Mary chews her pacifier *because* she is in Freud's oral biting stage of psychosexual development. John failed the IQ test *because* he is stupid. Emily loves to go to parties *because* she is gregarious. Amos keeps to himself *because* he is an introvert. I have italicized *because* in each of the examples to emphasize the nature of the basic problem. In my opinion, the concepts of negativistic stage, oral biting stage, intellectual inferiority,

[1]E. B. Holt, *Animal Drive and the Learning Process* (New York: Henry Holt, 1931), pp. 4–5.

[2]The diagnosis of "minimal brain damage" so lavishly used by those who do not know anything about the brain may be even more of a horror story. A particularly destructive consequence of such vacuous "explanations" is that they lead to inactivity. If a disorder is the result of "brain damage," possible remedial steps will not be taken on the assumption that nothing can be done about the problem.

gregariousness, and introversion all have some usefulness—but not as explanations. They only describe, and what they describe includes the thing to be explained. The explanation is circular.

The last point may suggest to you that, in order to explain a phenomenon, the important thing will be to break the circle and that this might be accomplished by relating the to-be-explained phenomenon to something else. This is an important insight: one definition of explanation is that it consists exactly in relating a phenomenon to some other set of conditions.[3] This brings us finally to the point of including this prologue. Graphs are a way of showing the relationship of something to something else. And since relationships are essential to explanation, graphical representation contributes in essential ways to the mission of science, the explanation of natural phenomena.

INDEPENDENT AND DEPENDENT VARIABLES

The previous section spoke of relationships between some to-be-explained phenomenon and some other set of conditions. In another way of speaking, the to-be-explained phenomena are *dependent variables* and the other sets of conditions are *independent variables*. In these terms the relationships to be discussed in this chapter are between independent and dependent variables; they show how some dependent variable changes with changes in some independent variable. It is time, now, to make these ideas more concrete. I shall do this with the aid of several familiar examples.

Cigarette Smoking and Lung Cancer. "Warning: The Surgeon General has determined that cigarette smoking is dangerous to your health."[4] This notice, which, by law, appears on all packages of cigarettes and in all advertising for cigarettes, actually states a relationship in rough form and will serve as a first example of the meaning of dependent and independent variables. It says that the danger of lung cancer and certain other illnesses (dependent variable) increases (relationship) with cigarette smoking (independent variable). For a good many purposes it is important to state relationships more precisely than that. For the moment, however, the statement above will do.

[3]There are two other definitions of explanation that one encounters with some frequency. One holds that the explanation of a phenomenon consists in showing that it can be derived deductively from a set of propositions. The other is that to explain a phenomenon is to account for the variation in it. The relationships among these definitions are far beyond what I wish to cover in this book. Later, however, I will have some things to say about accounting for variance.

[4]Just how dangerous is not entirely clear. Surely an editor of the *Colorado Daily*, a student newspaper, went too far when in a spasm of statistical imagination he announced that "The death rate for smokers is 70 per cent higher than for the rest of the population." For those of us who smoke I suppose that this makes our mortality rate 170%, a terrifying prospect indeed.

Scholastic Aptitude and School Grades. The basis for requiring scores on such tests as the SAT (Scholastic Aptitude Test) as a part of the materials a student must submit with an application for admission to many colleges is that there is a modest positive correlation between such scores and the student's later grade-point average. Generally, the higher the applicant's SAT score, the better her/his performance in college. In this case the translation into more general terms is that college grades (dependent variable) vary directly with (relationship) SAT scores (independent variable).

Intelligence and Preschool Experience. The SAT scores in the previous examples are measures of intelligence. There is good experimental evidence based upon other measures of intelligence that preschool attendance (independent variable) can increase (relationship) performance on an IQ test (dependent variable) by about 10 points on the average. This example will serve to make two points:

1. Whether a variable is dependent or independent cannot be determined from the nature of the variable. In the SAT example, a measure of intelligence was an independent variable; in the preschool example, a similar measure was a dependent variable. The determination of the status of variables depends upon whether the variable is a "predicted" variable (lung cancer, college grades, improvement in IQ) or a "predicted-from" variable (smoking, SAT scores, preschool attendance). *Predicted variables are dependent variables, predicted-from variables are independent variables.*

2. In experiments, independent variables are usually physical conditions that can be controlled and manipulated by the experimenter. Some writers limit the meaning of independent variable to those which can be manipulated in this way. Such a definition poses obvious problems for cases illustrated by our last two examples.

Dark Adaptation. When a person goes from the brightly lighted outdoors into a dark movie theater, he has trouble seeing at first. With time, however, vision improves. This improvement is dark adaptation. Experiments have shown that, with increasing amounts of time in the dark (independent variable), the intensity of the weakest light a person can see (dependent variable) decreases (relationship) for a period of 30 minutes or so.

This example again serves to show that the kind of variable does not tell you whether it is dependent or independent. In this case, the *intensity of a light*, a physical variable, is the dependent one in an experiment. Usually the physical variables in experiments are independent, manipulated variables, things such as numbers of trials, physical amounts of reward, and intensities of lights. As we have just seen, however, physical variables can be dependent. It depends upon the particular role the variable plays.

COORDINATES

Imagine that a market analyst has certain data on 10 companies listed on the New York Stock Exchange. The data presented in Table 2–1 are fictitious and are contrived to make points about the pictorial representation of data rather than economic points. Imagine, however, that for each company they are measures of the change in the profit picture from five years ago to the present and the change in the price of a share of stock over the same time span. The first measure is a percentage of change; the second is in dollars.

If you inspect these data, you will see that there is a relationship between the numbers in the two columns: profit increases are associated with increases in the price of a share of stock. The greater the increase or decrease in profit, the greater the increase or decrease in the price per share. A graph can make these points efficiently. It can also show a degree of regularity in the data that is lost in tabular presentation. The graph in Figure 2–1 does this and also provides a way of showing how graphs are made.

The construction of a graph begins with the laying out of *coordinates*, the horizontal and vertical axes of the graph. With this general point in mind it will be important, also, to note these more specific points.

1. If you will look back at the table, you will see that the percentage changes in profit go from a loss of 40% (−40) to a gain of 50% (+50). Now if you will look at the horizontal line (called the *X axis* or *abscissa*) in Figure 2–1 you will see that the scale on that line runs from a little less than −40 to a little more than +50. The extension of the scale beyond the range of the numbers to be dealt with is for esthetic purposes and nothing else.

TABLE 2–1
Prices and Profits on the Stock Market

Company	Five-Year Percent Change in Profit	Five-Year Dollar Change in Price of a Share
A	−10	−2
B	+20	+4
C	+40	+8
D	0	0
E	+10	+2
F	+30	+6
G	−40	−8
H	+50	+10
I	−20	−4
J	−30	−6

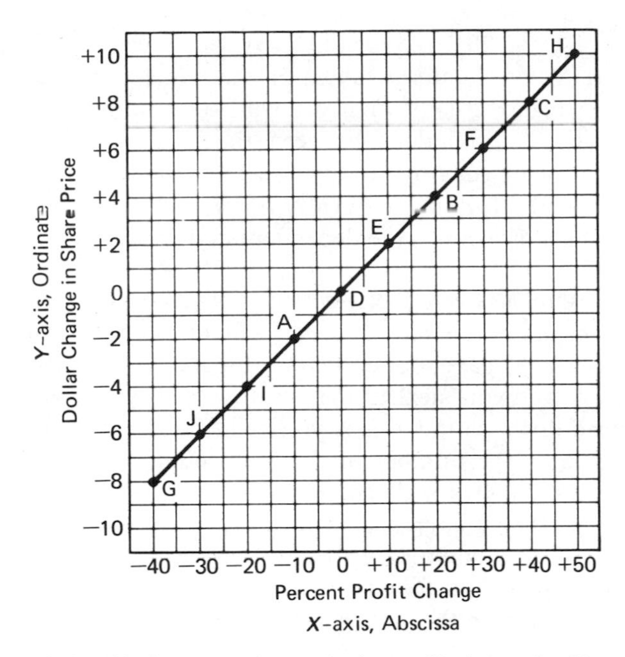

Figure 2–1 Relationship between change in the profit picture for 10 companies and the changes in the price of a share of stock. The numbers were made up; nothing as complex as the relationship involved ever turns out to be that neat. The resulting graph is an example of an increasing linear function. Note that any company whose change in profits was 10% better than another company increased in selling price by $2 more than that other company (or experienced a *decrease* of $2 *less*, which amounts to the same thing).

2. The changes in price per share go from a -8 dollars to a $+10$ dollars. The vertical axis (called the *Y axis* or *ordinate*) in Figure 2–1 covers the same range.

Probably the question you want to ask at this point is why I put percentage profit change on the *X* axis and dollar change in the price of a share on the *Y* axis instead of vice versa. The answer is that the decision in this case is somewhat arbitrary. *In general, values of an independent variable go on the X axis of the graph and values of a dependent variable go on the Y axis*. In many cases independent variables are more like causes, and I tend to think that the price per share of a stock would go up or down *because* profits of a company go up or down instead of the other way around. Thus the profit change in my thinking is the independent variable, and the change in price per share is the dependent variable. This accounts for how I decided to plot the data.

There are cases where the decision would be almost totally arbitrary. Graphs like Figure 2–1 could be made up to show the relationship between height and weight of 20 five-year-old boys, or grades for 50 students in sociology and history courses taken in the same semester. These are examples (as is Figure 2–1) of what you will come to understand as *scatter plots*, which depict correlations between the measures obtained on the same individual or object (pp. 161–78). In the case of correlations, statements of cause and effect are frequently nonsensical and there is no basis for deciding which measure is the dependent and which is the independent variable.

The relationship between profit change and change in price per share for our 10 imaginary companies displays itself when one measure is plotted against the other. I have done this in Figure 2–1, leaving the letter identifying each company next to the point for it. If you take one of these points and read down to the *X* axis you will find that the value on the *X* axis is the same as the number in the first column in the table for the company in question. If, for the same point, you read over to the *Y* axis, you will find that the number is the same as it was in the second column of the table. This should come as no great revelation because that is where the numbers came from in the first place. If you failed to get the numbers back, you would be sure that the graph was drawn incorrectly.

TYPES OF RELATIONSHIP

Having presented the basic ideas involved in making graphs I would like first to give you a good bit more practice in interpreting graphs, and second, to describe some of the most important relationships they can portray.

To begin the discussion, it will be useful to note that one message a graph can convey is that there is no relationship between two variables. Consider, for example, the number of hours an "unemployed" housewife devotes to household work and how things have changed over the years. Data happen to be available for a selection of years between 1927 and 1966.[5] They are presented graphically in Figure 2–2. The base line (*X* axis, abscissa, independent variable) represents a selection of years from 1927 to 1966. Please note that the years are not equally separated in time and that their locations on the *X* axis are separated by unequal distances for that reason. The vertical axis (*Y* axis, ordinate, dependent variable) is the average number of hours per week the housewife devotes to housework. In this case, please note the break in the vertical axis, which is there to indicate that hours of work from about 10 to about 40 have been left out. This is simply for purposes of making the graph somewhat more attractive.

[5]J. Vanek, Time spent in housework, *Scientific American*, **231**, 1974, pp. 116–120.

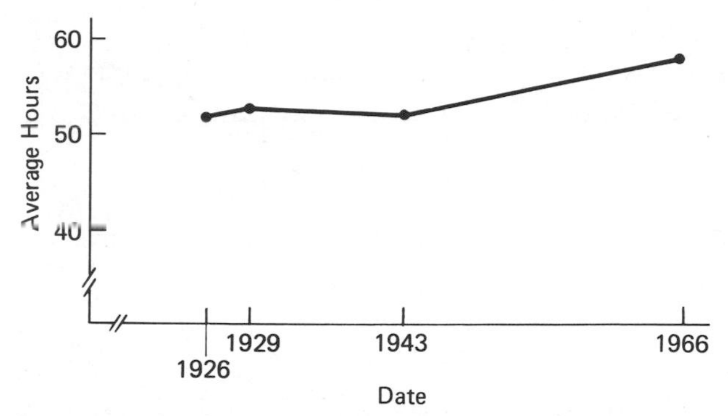

Figure 2–2 Average number of hours per week a housewife devotes to housework in selected years from 1926 to 1966. The impressive thing is that the number of hours has remained stable during this 40-year period. This is in spite of the fact that many labor-saving developments took place during the same period.

The facts presented by the graph are not very attractive. On the average a housewife devotes more than 50 hours a week to household tasks. And, despite the appearance of the electric washer, automatic dryer, wash-and-wear clothing, automatic dishwasher, self-service shopping, and convenience foods during the period covered, the number of hours per week has remained remarkably constant. Although the graph shows a trend that appears to be very slightly upward, the important observation to make is that over the period of 40 years represented there has been little or no change in the number of hours the average housewife devotes to household chores. Statistical tests show that the differences are not significant (the null hypothesis cannot be rejected).

How to Be a Graphic Liar

Suppose that the data in Figure 2–2, on the length of the average housewife's work week, were to find their way into print in newspapers with three different levels of respect for the truth, revealed by their headlines.

Conservative and Factual
Housewives' Work Week Constant
Little Change in 40 Years

Mildly Sensational
Women's Work Is Never Done
Labor-"Saving" Gadgets Add to Labor

Tabloid Scare
WIVES' WOES WORSEN
American Family in Grave Danger

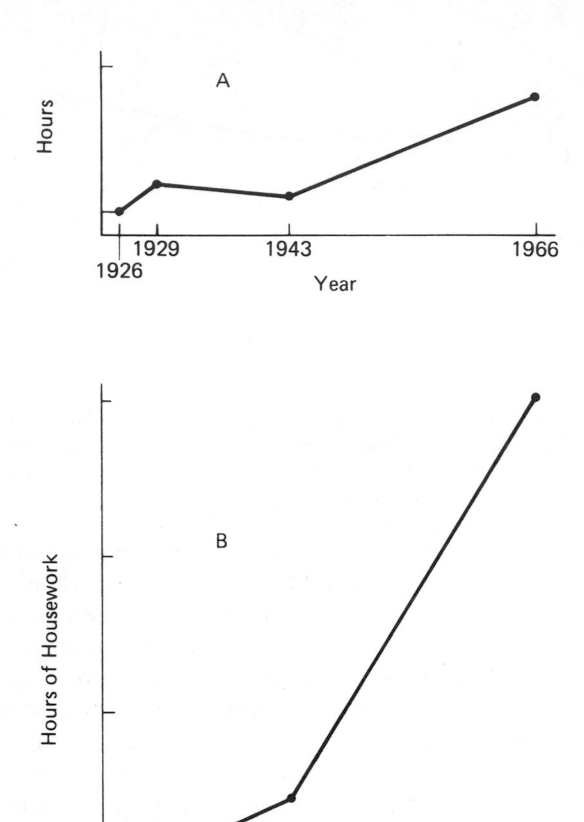

Figure 2–3 Exaggerating with graphs. Actually, there is no statistically significant increase in the number of hours per week the average housewife works. Various tricks in the drawing of graphs can create the impression that there is, however.

You can fill in the increasingly imaginative prose for yourselves. The first account tells the story as it is. If it wanted to illustrate the trend, Figure 2–2 would do very well. For more sensational purposes there are several ways to make the increase in working hours seem more substantial. Figures 2–3A and 2–3B illustrate some of them.

1. *Eliminate zero and shorten the Y axis.* Figure 2–3A shows this effect. The plot is exactly the same as in Figure 2–2 but only a small slice of the graph appears there. Since the plot goes from almost the bottom of the graph to almost the top, this exaggerates the effect.

2. *Eliminate numbers on the Y axis.* In order to create the impression of even greater change, it would be possible simply to omit the values on the ordinate in Figure 2–3A, leaving the reader with only a perceptual impression to go by.

3. *Expand the Y axis.* This effect is shown in Figure 2–3B, which also illustrates the usefulness of another gimmick.

4. *Select data to make your point.* In Figure 2–3B I have omitted the data for 1929, which for some purposes would only produce a distracting irregularity.

Positive and Negative Functions

Figures 2–1 and 2–3B, whatever their other deficiencies, are good examples of *positive functions*: increasing values along the *X* axis of the graph are associated with increasing values on the *Y* axis. Graphs where increases on the *X* axis are associated with *decreases* on the *Y* axis are *negative functions*. The recent downward trend in birth rates supplies an example. The 1976 *World Almanac* (p. 959) reports the data on birth rates shown in Table 2–2. Birth rates are the number of live births per thousand of the population. Before you read further, I suggest that you examine the table because there is one number in it that should make you wonder. It is the birth rate for 1972. In a period when birth rates are showing a steady decline, the rate for 1972 appears to be the highest in recent history. Why would that be? Before you spend too much time creating fanciful theories (e.g., blaming it on blackouts resulting from power failures), I suggest that you consider a simpler explanation—that the number is a misprint. If one examines the rest of the table in the almanac, it quickly becomes clear that that is all it is: the number should be 15.6.

TABLE 2–2
Birth Rates by Year

Year	Birth Rate
1955	24.6
1960	23.7
1965	19.4
1970	18.4
1971	17.2
1972	25.6
1973	14.9
1974	15.0

The data, corrected for that error, are plotted in Figure 2–4. The abscissa (*X* axis) is time again; the ordinate (*Y* axis) is birth rate. In that form the data give us an example of a *negative function*. Increases in *X* (years) are associated with decreases in *Y* (birth rate).

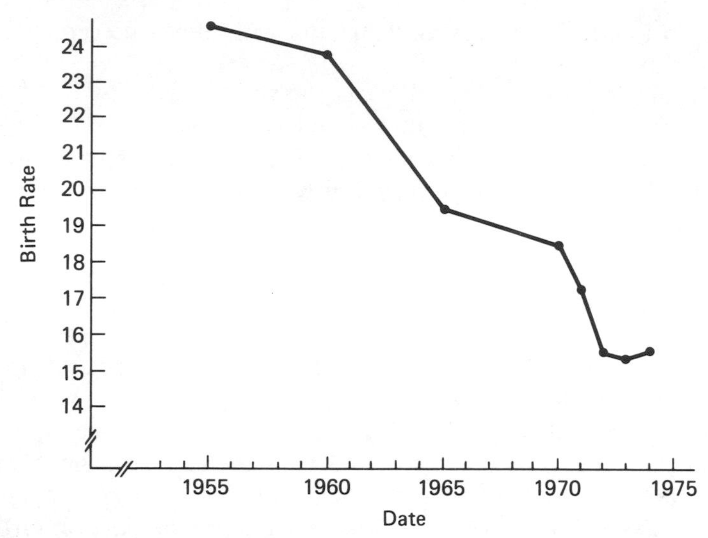

Figure 2–4 Birth rate over the years. This is an example of a negative function. The birth rate has been decreasing fairly steadily. The last data plotted (1973 and 1974) suggest that the function may have reached a lower limit. Data for more recent years indicate that the trend has, in fact, leveled off.

Linear and Curvilinear Functions

Figure 2–1, as mentioned in the figure caption, is a *linear function*. This means that equal changes on the X axis are always associated with the same amount of change on the Y axis. Figure 2–5 presents a set of four linear functions, two positive and two negative, which make this point more generally. A little study of these figures will show two things: (1) for each function individually, the amount of change in Y is always the same for the same change in X (+2.5, +5.0, −1.5, and −2.4 in the four graphs), and (2) as these numbers show, the amount of change for a given change in X is different for the different functions. The technical way to say this is that the different functions differ in *slope*. The greater the change in Y for a constant change in X, the greater or steeper the slope.

In a linear function, the rate of change is constant: equal increments in X, wherever they occur, are associated with equal increments in Y. Functions in which equal increments in X are associated with changing increments or decrements in Y are not linear but *curvilinear*. In graphic form, they are not straight lines as linear functions are, but curved lines.

Increasing and Decreasing Returns

Probably everyone is familiar with the story of the peasant who saved the King from highwaymen. As a reward, the King offered the peasant a position at court and asked the peasant to set his own wages. "Oh, sire,"

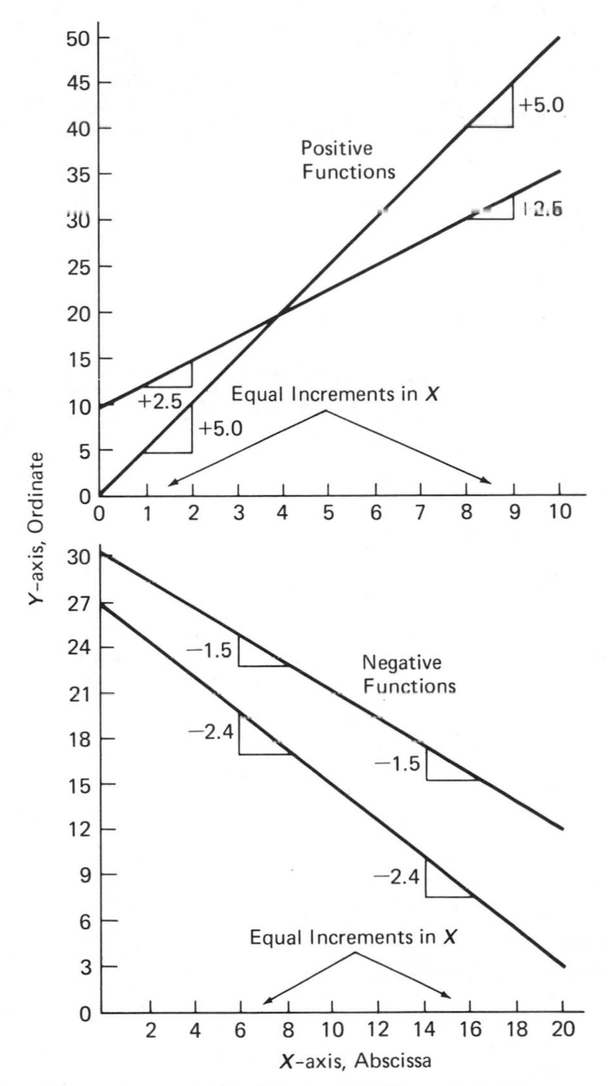

Figure 2–5 Positive and negative functions with different slopes.

the peasant said, "I really want very little. Employ me if you so desire, Majesty, but only for a month. And for my wages pay me just a penny for the first day, two for the second, four for the third, and so on for 30 days. I am worthy of no more."

If the unwary King accepted this offer, he made a very bad bargain. Figure 2–6 shows what happens to the peasant's wages just for the first 11 days. From the plot you can see that the rate of pay is beginning to rise rapidly. You can also see why I plotted the peasants wages for only 11 days. Day by day they increase by an increasing amount—only 1 cent

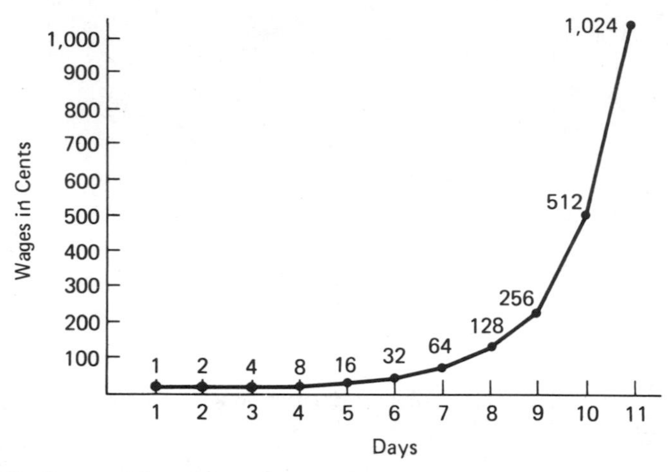

Figure 2–6 Curve of increasing returns. The numbers in the curve are the values plotted. I have entered them so that, by subtraction, you can get a better picture of the meaning of increasing returns.

from day 1 to day 2, but 512 cents from day 10 to day 11. From day 29 to day 30 the increase is from about 311 million cents ($3.11 million), to about 622 million cents ($6.22 million). This is what I shall call a *curve of increasing returns*, in which uniform increments in X are associated with increasing increments of Y. As an aid to check on this point, I have left the values of Y at the appropriate points on the curve in Figure 2–6. If you subtract the smaller from the larger of adjacent values, you will see that these differences get larger and larger as the number of days worked (X) increases by even amounts.

Curves where changes in Y become less and less as X increases are *curves of decreasing returns*. To illustrate this type of function, and some other points as well, data on crime rate as a function of the size of a community will be useful. Table 2–3 presents the data.

TABLE 2–3
City Size and Crime

Average Size of City	Crimes per 1,000 Population
750,000	65
145,000	62
70,000	49
35,000	44
16,000	38
4,500	33
Rural	15

Obviously, the crime rate increases with city size, but exactly what the relationship is like is hard to visualize because of the unequal spacing of sizes of cities. A graph makes things much neater, as Figure 2–7 will show. The curve is a curve of diminishing returns of the type under discussion. As the size of the city increases, crime rate increases but much less dramatically when one compares cities of 145,000 to those of 750,000 than when the comparison is between cities of 4,500 and 16,000, for example.

Curves of the type shown in Figure 2–7 appear to be going toward some upper limit which they would never exceed. In the specific case in Figure 2–7, the idea would be that the crime rate would not increase very much with larger and larger cities, that the rate of 65 per thousand for cities of 750,000 is not far short of what it would be for cities of 1 million or even 10 million. If you look back at Figure 2–4, you will notice that it appears to be approaching a lower limit which it would not go below. Such upper and lower limits which functions approach but do not exceed are technically called *asymptotes*. The concept of asymptote will have some significance in later discussions of statistical functions.

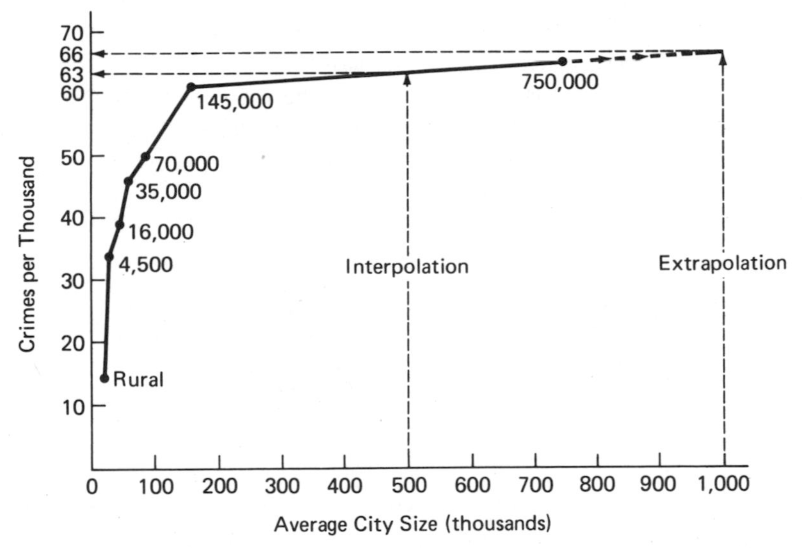

Figure 2–7 City size and crime rate. The function is a positive function with decreasing returns. This graph also aids in understanding the concepts of *interpolation* and *extrapolation*. Crime rates for cities of half a million population do not appear in the figure. You might estimate the rate at about 63/1,000, however, by *interpolating* a point on the curve by reading up from the abscissa, from 500, 000 to the curve, and then reading over to the ordinate. Cities of a million are not represented either. If you extend the last segment of the curve by continuing the straight line, it can be *extrapolated* to 1,000,000. Then reading over to the ordinate, you could estimate a rate of about 66/1,000. The dashed arrows in the curve illustrate the process.

This curve also makes it easy to describe the concepts of *interpolation* and *extrapolation* which sometimes come up in connection with graphs. Both of these refer to ways of estimating data for points not represented in the graph. In Figure 2–7, for example, there are no data between city sizes of 145,000 and 750,000 or beyond 750,000. You could estimate crime rates for intermediate-sized cities by *interpolation*, that is, by reading from the curve at the location of the desired size of city to the ordinate. You could estimate the crime rates of larger cities by extending the curve to the desired size and again referring to the ordinate. This process is called *extrapolation*. Figure 2–7 provides examples.

Ogives

A type of curve that will have some importance in our later discussions is *ogival* in shape, or S-shaped. These occur with considerable frequency as growth curves which show the relationship between age and some physical or psychological characteristic. Figure 2–8 shows such a curve for the development of passive vocabulary, the number of words a person can understand. It begins, of course, at zero, shows a period of increasing returns. Finally, there is a period of decreasing returns, leading to a symptotic understanding of some 60,000 words.

Figure 2–9 is a negative ogive. It shows the proportion of men, all of whom were alive at age 30, who still survive at various ages. Again the curve begins with increasing returns and ends with a period of decreasing returns, reaching zero at age 110.

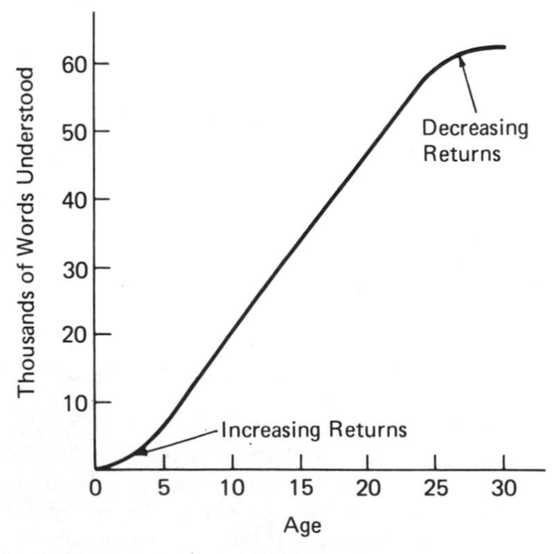

Figure 2–8 Growth of passive vocabulary. This curve shows an early section with increasing returns, and a later section with decreasing returns. In between, the curve is essentially linear.

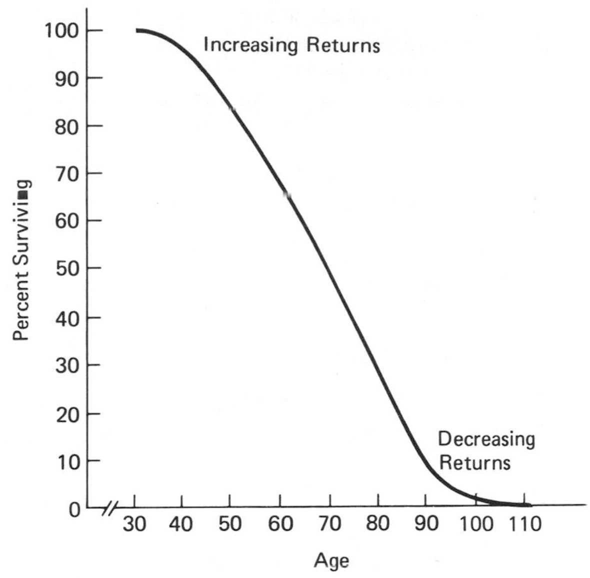

Figure 2–9 Negative ogive.

Monotonic and Nonmonotonic Functions

All the relationships presented so far in this chapter have been *monotonic*: as X increases, Y increases or decreases and never changes direction. A curve that does change is *nonmonotonic*. The following discussion leads up to an example.

People who have done it say that parachute jumping is one of the most exhilarating, but also one of the scariest experiences a person can have —especially the first jump: The experience of the first jump remains one to be talked about for many months to come. But only about 15% of the people who make a first jump ever take a second jump. In the words of one of these dropouts, "It was one of the most exciting experiences in my life, but I sure would never, never ever want to go through it again."[6] Studies of novice jumpers about to take their second jump tell us a great deal about the frames of mind of these people. Physiological measures show that they are in a state of great arousal. Tests of their hearing show that it is slightly impaired. Word-association tests show that they are obsessed with thoughts of parachuting. Table 2–4 presents the responses of one jumper asked to give "the first word that comes to mind" for each of the stimulus words. Asked to make up stories about pictures, the jumpers again reveal that they are preoccupied with jumping. Some stories showed that these people are afraid of making the jump; others show that they are afraid but deny their fear.

[6]W. D. Fenz, Conflict and stress as related physiological motivation and sensory, perceptual and cognitive functioning, *Psychological Monographs*, **78**, 1964, Whole No. 8.

TABLE 2–4
Word Associations of Novice Parachutists

Stimulus Word	Response Word
Skydiver	Jump
Hungry	Jump
Radiator	Hot
Happy	Parachute
Bicycle	Jump, damn it, I can't get away from it!

Our concern here will be with some self ratings that beginning parachutists made of their feelings at several times ranging from a week before the jump until just after landing. The ratings were on two scales. One was a scale of positive anticipation (approach) defined for the jumpers as "looking forward to the jump, wanting to go ahead, being thrilled by the prospect of jumping." The other was a scale avoidance, "wanting to turn back and call the jump off, questioning why you ever got yourself into jumping, fear." The ratings on both scales were from 1 (weakest) to 10 (strongest). The averages of these ratings are shown in Figure 2–10. Both

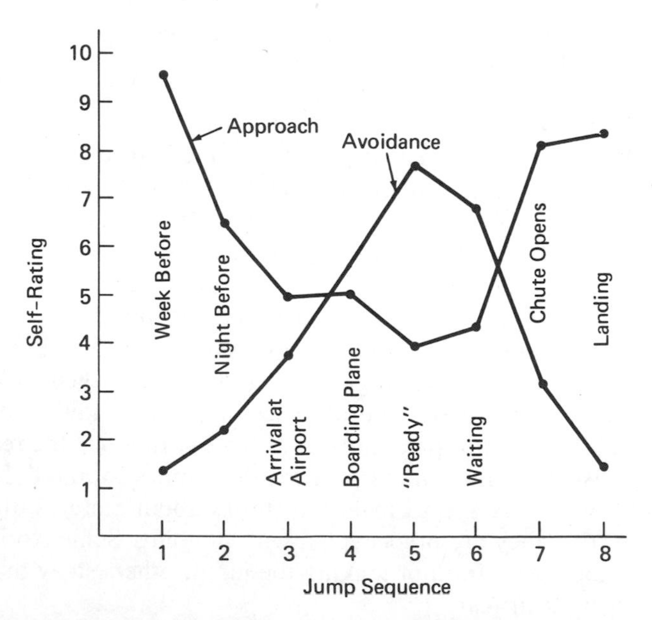

Figure 2–10 Nonmonotonic functions. The characteristic of nonmonotonic functions is that they change directions, at least once. In these cases the curve of approach decreases and then increases; the curve of avoidance decreases and then increases.

curves are *nonmonotonic*. The one for approach starts high, decreases until the jump occurs, and then rises again. The curve of avoidance starts low, rises until the jump occurs, and then decreases. Both curves change direction and in this way differ from all the curves discussed earlier in this chapter.

SUMMARY–GLOSSARY

This presentation could go on, but probably not productively. Figure 2–11 provides a graphic summary of this chapter on graphs. The important items for review are covered in the following glossary.

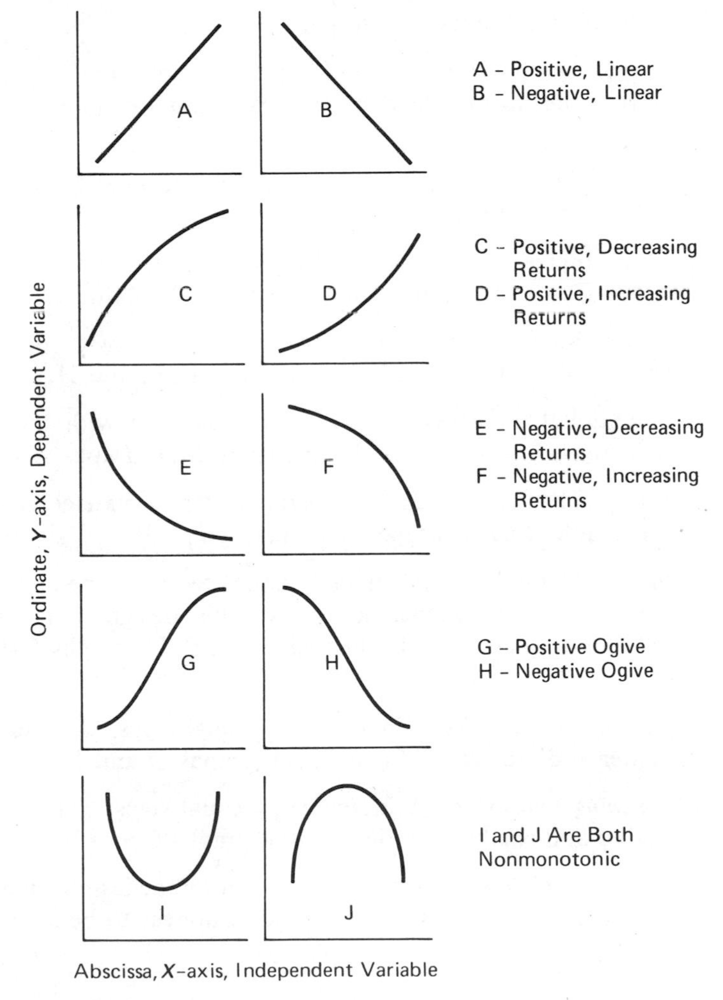

Figure 2–11 Summary of some types of graphs.

Explanation. As used in this chapter, relating a phenomenon of interest to something else. Elsewhere, (1) showing that a phenomenon follows deductively from a set of assumptions, or (2) accounting for the variance in a phenomenon.

Circular "explanation." "Explaining" by naming. Occurs when the phenomenon to be explained and the "explanation" are the same thing.

Dependent variable. A phenomenon of interest or the object of study in some realm of science.

Independent variable. A variable to which a dependent variable is related, or which varies in a study to determine whether such a relationship exists.

Coordinates. The X and Y axes of a graph.

Abscissa (X axis). The horizontal axis of a graph. The independent variable in the relationship presented by the graph usually appears on the abscissa.

Ordinate (Y axis). The vertical axis of a graph. It usually carries a representation of the dependent variable.

Positive function. Increases in X area always associated with increases (never decreases) in Y. In Figure 2–11, graphs A, C, D, and G.

Negative function. Increases in X are always associated with decreases (never increases) in Y. In Figure 2–11, graphs B, E, F, and H.

Linear function. Equal increases in X always associated with the same amount of increase or decrease in Y. In Figure 2–11, graphs A and B.

Slope. The rate of change of a curve. In graphic terms, its steepness. In Figure 2–11, graph A has a steeper slope than graph B.

Curvilinear function. Equal increments of X associated with unequal increments in Y. It is conventional to reserve this designation for the monotonic functions (see below). Thus, in Figure 2–11, graphs C–H are curvilinear.

Curve of increasing returns. As X increases by equal steps, changes in Y become greater and greater. In Figure 2–11, graphs D and F.

Curve of diminishing returns. As X increases by equal steps, changes in Y become smaller and smaller. In Figure 2–11, graphs C and E.

Asymptote. An upper or lower limit that a curve approaches but does not exceed. In Figure 2–11, graphs C, E, G, and H appear to be approaching asymptotes.

Interpolation. Estimating Y for a value of X for which a point is not plotted but where X is within the range of values of X appearing in the graph. To interpolate is to place a point between points plotted in a graph.

Extrapolation. Estimating Y for a value of X that is smaller or larger than any value of X represented in the graph. To extrapolate is to extend the graph.

Ogive. An S-shaped curve showing sections of increasing and decreasing returns. In Figure 2–11, graphs G and H.

Monotonic function. A function where increases in X are everywhere associated with increases or decreases in Y. In Figure 2–11, graphs A–H.

Nonmonotonic function. A curve that changes general direction. In Figure 2–11, graphs I and J.

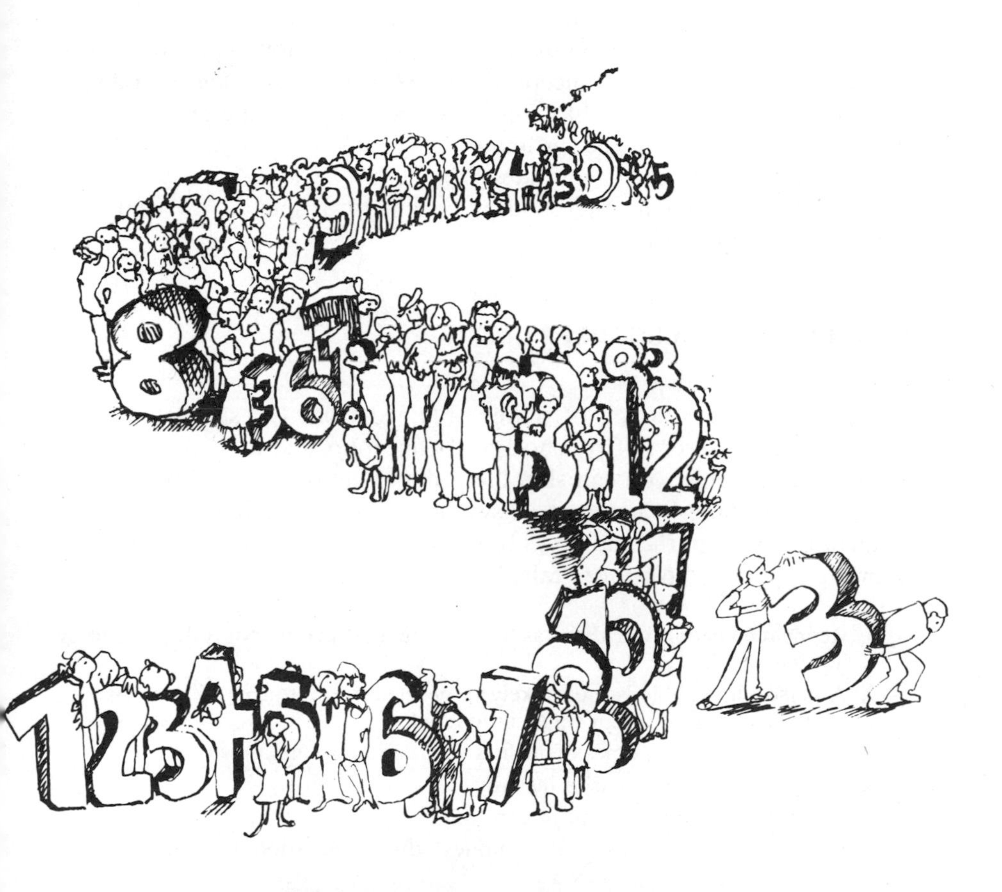

Frequency Distributions

In statistics by far the most important function is a nonmonotonic one, the normal distribution. I want to put off discussion of it, however, until I have presented the more general concept of the *frequency distribution*. Frequency distributions are graphs in which the abscissa represents the magnitude of some phenomenon and the ordinate represents the frequency of occurrence of each of these magnitudes. The definition I have just given you is a good bit harder than the concept I want to define. A few examples should clear things up.

PSYCHOLOGICAL MISCONCEPTIONS

Figure 3–1 is a distribution of scores obtained from 122 students on the following true–false test of psychological conceptions and misconceptions. If you are in a hurry, you can skip the test and get on with the discussion. If you want to take it, however, answers and an occasional comment on the answer appear in the end notes for this chapter. Just answer each of the following questions "true" or "false."

1. Redheaded people are about as likely to be mild-mannered as they are to be hot-tempered.
2. A person in trouble is more likely to get help from a group of witnesses to his problem than from a single individual.
3. The causes of most types of mental retardation are unknown.
4. Genius is closely related to insanity.
5. Preacher's children are apt to turn out badly.
6. Phi Beta Kappas make more money after graduation than the average college student.
7. Watching examples of aggressive behavior (for example, on TV) lessens a child's tendency to be aggressive.
8. Mental patients are usually not dangerous to other people.
9. The exceptionally intelligent person is apt to be as strong physically and as well-adjusted socially as the average individual.
10. Even in the most masculine and the most feminine of us there are traces of the opposite sex both physiologically and psychologically.

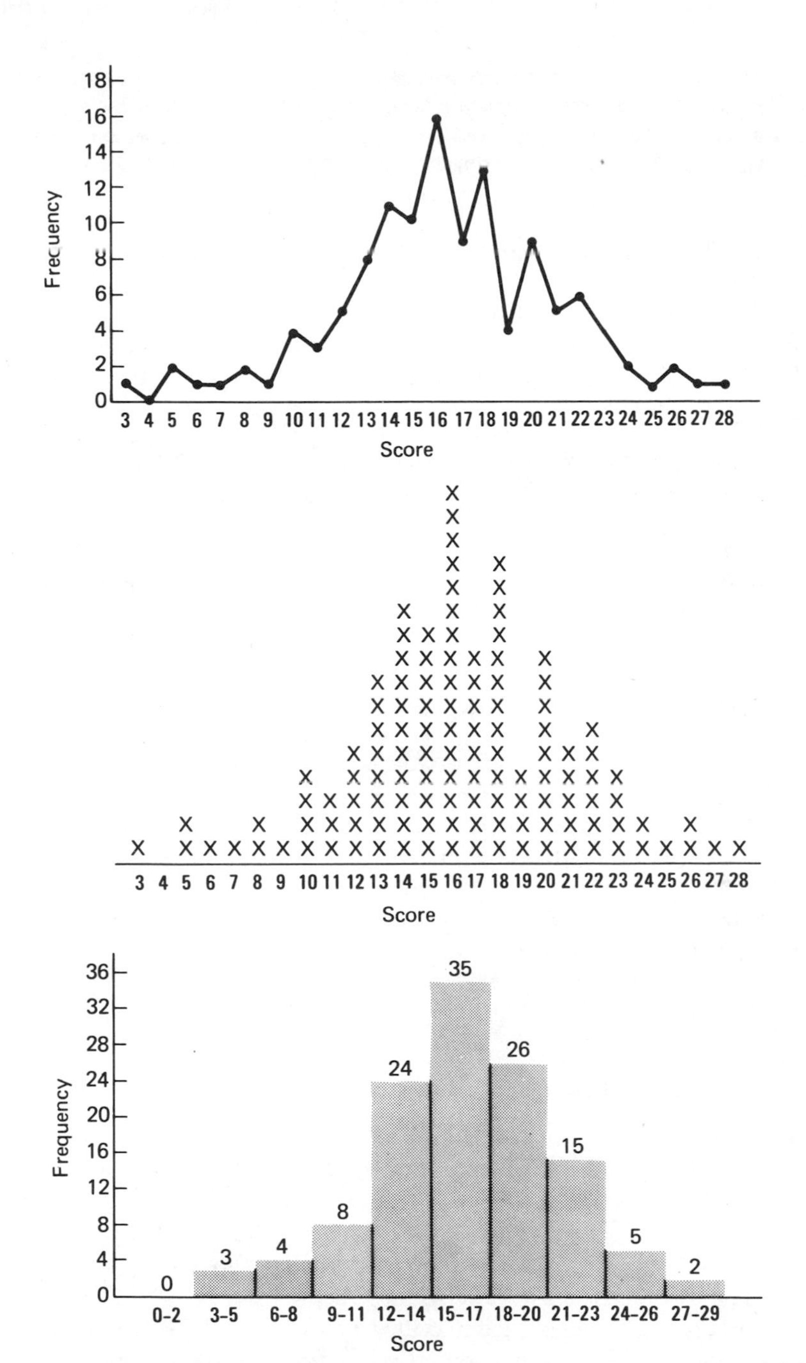

Figure 3–1 Types of frequency distribution. The upper graph is a frequency polygon. The middle graph shows where each individual fell in the distribution above. The bottom figure is a bar graph or histogram combining data by class intervals of 3.0. In a graph such as this one, *N* refers to the number of cases represented in the graph.

11. In selecting people for a job it is better to rely on letters of reference from people who know the applicant well than on tests of aptitude.
12. There is a law of compensation in nature. For example, a person born blind is also born with a highly developed sense of touch.
13. Public opinion polls are frequently accurate within one or two percentage points.
14. People, including parents, are not very good at judging the emotions of babies from the infants' behavior.
15. The mixture of red and green lights produces yellow.
16. The power of subliminal advertising is now well established.
17. It is seldom possible to prove that a mental breakdown occurred because of a single experience.
18. If a person becomes schizophrenic, it is probably more because of how the mother treated him/her than because of any genetic factor.
19. The methods used in "brainwashing" are those first studied by the Russian physiologist Ivan Petrovich Pavlov.
20. An especially favorable environment can raise a child's IQ by a few points.
21. The most successful women in what is still pretty much a man's world are most often those with older brothers.
22. As far as friendships are concerned, opposites tend to attract.
23. People have better vision than white rats.
24. Obese people tend to be overly sensitive to the internal stimuli that provide signals of hunger.
25. Sleep and hypnosis are two very different states of consciousness.
26. Babies' earliest smiles are for their mothers.
27. If a person threatens suicide, he almost certainly will not attempt it.
28. A neurotic person stands a 50–50 chance of losing the neurosis in 5 years whether or not he receives therapy.
29. There is some tendency for people to postpone death until after a birthday.
30. You are apt to forget more in the first few minutes after learning something than in the next several hours.

TYPES OF FREQUENCY DISTRIBUTION

The uppermost graph in Figure 3–1 is the most typical form of frequency distribution, a *frequency polygon*. The base line is the range of scores actually obtained on the test from a low score of 3 to a high score of 28. The ordinate shows the number of individuals who have obtained each score. The graph itself shows how these scores were distributed. If what this means is at all unclear, the middle graph in Figure 3–1 should aid your understanding. There I have entered an × for each of the 122 individuals represented in these data. You could reproduce the upper figure by connecting the tops of these columns of ×s with straight lines.

The bottom graph in Figure 3–1 is a *bar graph*, or *histogram*, in which I have presented the data in terms of *class intervals* which combine frequencies for each 3-point range of scores. The numbers at the bottom of each bar are the numbers of individuals obtaining scores in each of the ranges. You could check these entries by counting ×s for the indicated ranges in the middle figure. You could also convert the middle figure to a different bar graph by drawing rectangles around each column of ×s.

An important feature of the data in Figure 3–1 is that, however they are presented, the distribution is roughly symmetrical. They pile up in the middle range of scores and roughly equal numbers of scores occur on either side of this central tendency.

Cardinal Virtues and Deadly Sins

Have you noticed that things that come in numbers tend to come in threes or sevens? All Gaul, according to Caesar, was divided into three parts; the holy trinity consists of Father, Son, and Holy Ghost. Freud invented the concepts of id, ego, and superego to describe human personality and unconscious, preconscious, and conscious to identify different levels of awareness. In the realm of sevens there are seven wonders of the ancient world, seven ages of man, seven hills of Rome, seven deadly sins, and seven virtues. Actually virtues get into the act twice because three of them—hope, faith, and charity—are the cardinal virtues.

The origins of this interesting phenomenon are lost in history. One theory is that the emphasis on threes comes from the fact that, in ancient times, three was considered a nonconspiratorial number, the number of individuals who would not create a plot against another. This may of course be the effect of the tendency under discussion rather than the cause of it.

The basis for the magical power of seven is sometimes said to be the fact that a single circle can be surrounded exactly by six others of the same size, making a total of seven. You can prove this to yourself with seven coins. Again, however, there is no way of knowing whether the suggested connection is a real one.

However that may be, the record shows that the language does tend to preserve the memory of things that come in threes and sevens. Reference to an encyclopedia of trivia yielded the information presented in Figure 3–2,[1] where the tendency being described is clear. References to things that come in threes and sevens are more than twice as frequent as to things that come in any other number.

[1]The actual materials appear in the End Notes.

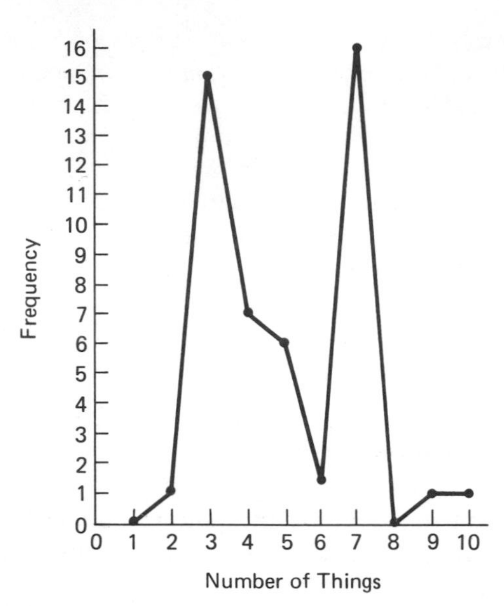

Figure 3–2 The numbers of things. Things that come in numbers show a pronounced tendency to come in threes or sevens.

A frequency distribution of the type shown in Figure 3–2 is called *bimodal* to contrast it with *unimodal* distributions such as those in Figure 3–1. A *mode* is a most frequent score in a distribution. Most distributions have one mode, which makes them unimodal. The distribution in Figure 3–2 has two, and thus is bimodal.

The Private Lives of Light Bulbs

Everybody knows that different samples of many products will vary in quality. Some automobiles really are "lemons." Figure 3–3 makes this point for light bulbs. The horizontal axis in this frequency distribution is the range actually obtained in tests of 150 bulbs. The concern in this study was with the length of time a bulb would last before it burned out. The only thing to note is that the axis is laid out in 200-hour intervals. The vertical axis is frequency, the number of bulbs that burned for each of the different numbers of hours.

This distribution is not *symmetrical*, it is *skewed*. It is conventional to call the two ends of the distribution where the frequencies are low the *tails* of the distribution. In this case, the longer tail of the distribution is toward the higher numbers. Such a distribution is said to be *positively skewed*.

Grades on tests in college courses often show the opposite picture. Most often because of upper limits on the highest possible score, scores pile up at the high end of the scale and "tail off" toward the lower scores. Such

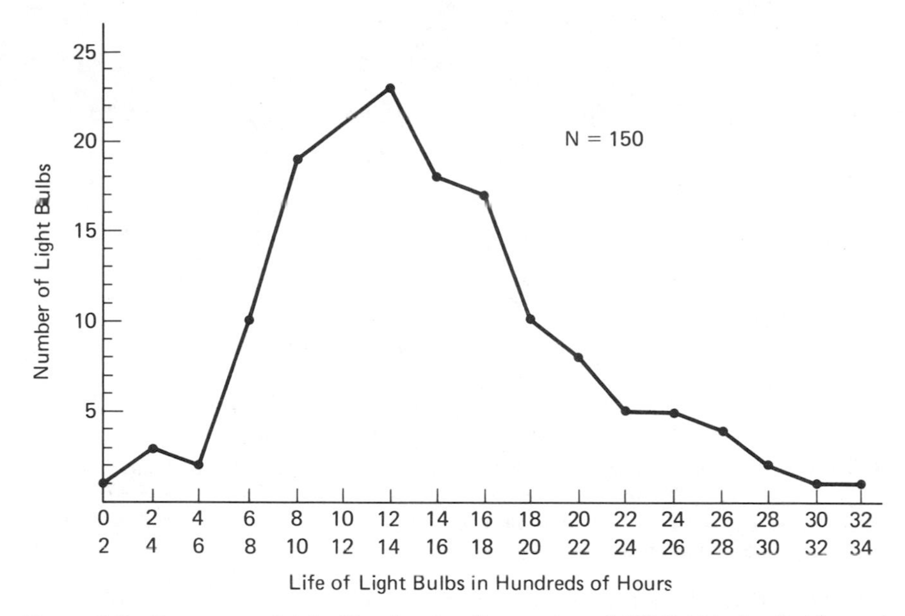

Figure 3–3 Frequency distribution showing the number of 150 light bulbs that burned for various number of hours from 200 to 3,400. Question: Where would you include a bulb that burned exactly 3,200 hours? Would it appear in the 3,000–3,200 or the 3,200–3,400 category? Answer: Such problems arise now and then. About all you need to do is to be consistent and include all such borderline cases in either the lower or higher category. This distribution is *positively skewed.* The long tail of the distribution is toward the higher numbers. *N* again is the total number of items represented, this time the number of light bulbs.

distributions are said to be *negatively skewed.* Whether a distribution is positively or negatively skewed depends on whether the long tail is toward the high numbers or the low ones.

Playing by the Rules

Most of us are conforming people. We keep off the grass, fasten our seatbelts, and extinguish all smoking materials when the stewardess tells us to do so, whisper or stay quiet in the library, and stop at stop signs when we are driving an automobile. But not absolutely everyone is so conforming. A frequency distribution of various types of behavior that motorists actually displayed at stop signs in one study appears in Figure 3–4.

For obvious reasons distributions of the type shown in Figure 3–4 are called *J curves*. They are obtained with some frequency in studies of conformity behavior, of which this would be an example. There are a couple of special points to make about this function: (1) the fact that the tail of the distribution is in the direction that would conventionally be

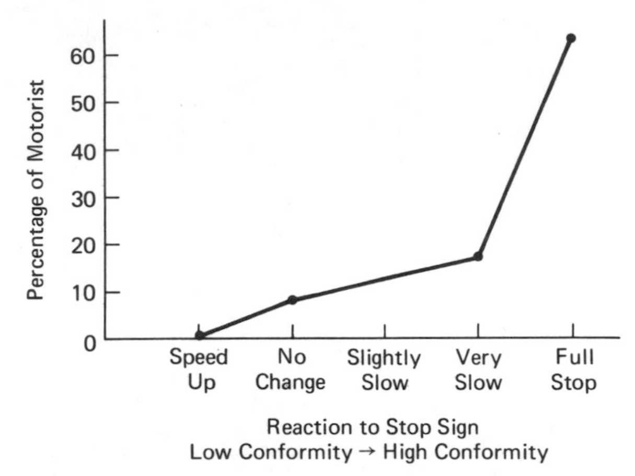

Figure 3–4 J curve. This curve plots the percentage of motorists behaving in various ways at stop signs. The function is typical of conforming behavior. I collected the data in Durham, N.C. A number of trips to Boston suggest that, there, the extreme left-hand value is probably not zero.

toward low numbers means that it is negatively skewed; and (2) frequency distributions can, as in this case, be made with percentages rather than raw frequencies on the ordinate.

HYPOTHESIS TESTING WITH FREQUENCY DISTRIBUTIONS

So far everything in this chapter has been about descriptive statistics. We saw in Chapter 1, however, that statistics also serve an inferential purpose. Your new familiarity with frequency distributions will make it a simple matter to illustrate this inferential function of statistics in a new way by *hypothesis testing*. I shall present two examples. In each case you will find that the treatment follows the same logic as I used to analyze the love life of Mr. Z (pp. 10–16).

The Federalist Papers

The time is 1787–1788 in the United States. The Revolutionary War is over and the country has won its freedom. A constitution has been framed and is before the people of the states to consider for ratification. In the state of New York a series of 85 federalist papers (published as *The Federalist*) appear, arguing in favor of ratification. The anonymous author, "Publius," is known sometimes to be John Jay, sometimes Alexander Hamilton, sometimes James Madison, and much less frequently a collaboration. The specific authorship, however, remained a mildly intriguing mystery.

Today the question of who wrote which of the federalist papers is a matter of more significance. The documents are important statements in the field of political philosophy. Historical evidence has served to identify the individual authors of 70 of the federalist papers. Three of the remainder are believed to be co-authored. For the remaining 12, however, both Hamilton and Madison claimed authorship. The question of who actually wrote them is harder to decide than you might think because of the circumstances under which the papers were written.

The positions taken in the various articles are of no help because the authors were functioning like lawyers arguing a case. The topics of discussion are of no help either, because they were assigned. General writing style is similarly nondiagnostic, because Hamilton and Madison were both experts in the complex style that was fashionable in the eighteenth century. For example, both used very long sentences—34.5 words in the average sentence for Hamilton, 34.6 for Madison. What eventually does turn out to provide a lead, however, is the way in which Hamilton and Madison used the little words (while, by, from, upon, to, etc.) that are necessary in written communications. The two men differed markedly in the frequency with which they used such words and a comparison of the disputed papers with those of known authorship provides the clue.

Take, for example, the word *upon*. For Hamilton, Madison, and the disputed papers the numbers of times per thousand words that this one occurs are as shown in Table 3–1. The entries are the rates for 48 writings (not just from the federalist papers) known to be by Hamilton, 50 known to be by Madison, and the 12 disputed papers. Without calculation you can see at once that the distribution for the disputed papers is like that for Madison. The distributions of rates for the word "by" are shown in Figure 3–5. The distribution for the disputed papers again resembles that for Madison. Other distributions for other words make a strong case for the conclusion that Madison rather than Hamilton was the author of the disputed federalist papers.

TABLE 3–1
Number of Papers Using Indicated Rates
Rate per Thousand Words

	.4	.4–.8	.8–1.2	1.2–1.6	1.6–2.0	2–3	3–4	4–5	5–6	6–7	7–8
Hamilton			2	3	6	11	11	10	3	1	1
Madison	41	2	4	1	2						
Disputed	11			1							

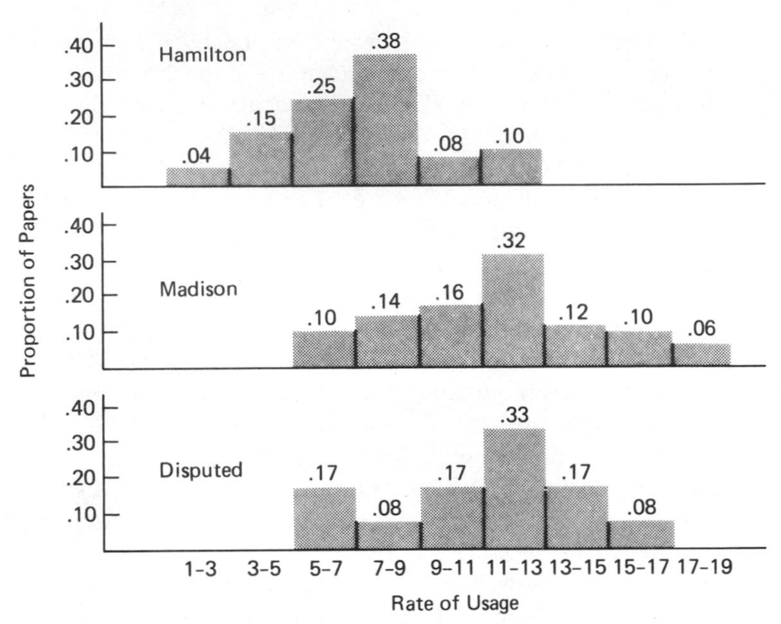

Figure 3–5 Distributions to show that the usage of the word "by" in the disputed *Federalist Papers* was more like Madison's usage than Hamilton's. This form of representation is called a *bar graph*, for reasons that are probably obvious. The proportion of usage is represented by the height of a bar. Again there is the problem of what to do about borderline rates (e.g., 7 per 1,000 words). In this case such rates were systematically included in the higher range (e.g., 7–9 rather than 5–7).

But if you accept this way of reasoning, note what you have done. Taking one of the two possible ways to describe it[2]:

1. You have stated a hypothesis in *null* form. There is no difference between rates of usage of little words for Hamilton and the author of the disputed papers.
2. You—really Mosteller and Wallace[3]—have collected the required data.
3. You have used your intuitive computer[4] to calculate a probability which says that it is highly unlikely that the distribution of rates for the disputed papers is the same as that for Hamilton.
4. You have rejected the null hypothesis and, by a process of elimination, you have come out in favor of Madison as the author of the disputed papers.

[2]Since I want to end up *rejecting* a null hypothesis, I have developed the argument in terms of a comparison of data for Hamilton and the disputed papers. The other way would be to make the comparison for Madison and the disputed papers, in which case you *accept* the null hypothesis.

[3]F. Mosteller and D. L. Wallace, Deciding authorship, in J. M. Tanur, et al., *Statistics: A Guide to the Unknown* (San Francisco: Holden-Day, 1972).

[4]This time the calculations do not fit into the organization of my computational appendix. They show, however, that the odds against Hamilton's authorship are 80 : 1 for the disputed paper where that conclusion is least secure. For the other papers, they are 800 : 1 and higher.

The logical steps are the same as for the subway puzzle in Chapter 1. There is, however, an instructive difference between this example and most actual applications of the methods of hypothesis testing. In this case rejecting the null hypothesis adds great strength to an alternative substantive hypothesis—that Madison wrote the disputed papers. This is because of the limited number of reasonable possibilities. Usually things don't work out that neatly. My final example will make that point.

Where Have All the Women Gone?

In 1968 Dr. Benjamin Spock came before the U.S. District Court in Boston to answer a charge of conspiracy to violate the Selective Service Act because he had encouraged resistance to the Vietnam war. Spock's lawyers challenged the legality of the jury selection procedure, which produced a jury with no women on it. This was a matter of serious concern because Spock's book on the care of babies had been the child-rearing bible of millions of mothers. The question was whether the all-male jury was the result of chance or of systematic discrimination against women. The answer to this question will take us through the logic of hypothesis testing two more times.

In order to understand the logic, it is necessary first to understand the methods by which juries are selected. The process involves a series of steps. In the first step the Clerk of the Court is supposed to select at random[5] the names of 300 potential jurors. In the second step a *venire* of approximately 30 people is drawn, again supposedly at random, from these 300. In the final step the 12 actual jurors are chosen from the venire at the beginning of the trial. Now for the statistical points. As the discussion develops, it is important to remember that it involves venires rather than jurors.

As is true generally, the proportion of women in the Boston population is slightly greater than the proportion of men. Thus except for the accidents of sampling, the 300 names selected by the Clerk initially should include the names of a few more than 150 women. Reasoning from the composition of the venires (see Figure 3–6), however, the evidence suggests that there were only about 90 women in the group. My own calculations indicate that this underrepresentation of women would happen by chance only once in several million samples of this type. Thus, following the logic of hypothesis testing (which I will not spell out again), I reject the hypothesis that the initial selection was by chance.

[5]More of this later (pp. 134–57). A *random sample* is one in which every individual and every combination of individuals has an equal chance of being drawn. If this is not the case, the sample is biased.

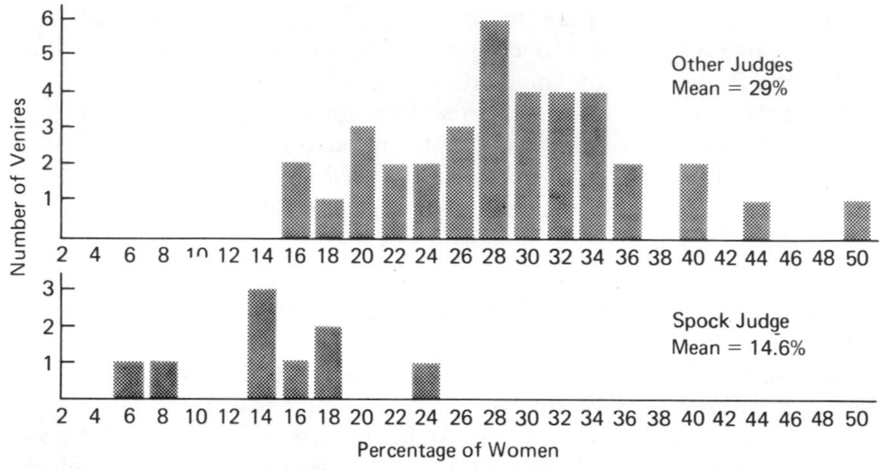

Figure 3–6 Proportions of women in venires of the Spock judge and other Boston judges. These graphs (again bar graphs) show that the Spock judge's venires tended to include fewer women than those of the other judges.

How about a jury with no women drawn at random from this already biased selection of 300 names, 210 men and 90 women? In this case it is instructive to look at the distributions of proportions of women in the venires of the Spock judge and the other judges in the same district. Figure 3–6 displays the data in the form of frequency distributions. There are 9 venires for the Spock judge and 37 for the others.

Obviously the distributions are different. The mean (average) percentage of women for the venires of the other judges is 29%, for the Spock judge 14.6%, and none of the latter venires has as many women as the average of the former. The intuitive conclusion urged by Figure 3–6 is confirmed by calculation. The odds against the venires of the Spock judge being a random selection from the Boston population are astronomical—1 in 1,000,000,000,000,000,000.

So where did all the women go? The question takes us back to the last point I made in the preceding section. In contrast to the Hamilton–Madison example, there is no neat alternative to consider. All we can say is that something introduced a bias against women at two stages in the process of jury selection, first in the selection of the 300 potential jurors and second in the selection of the Spock judge venires.

A BARGAIN AT SOTHEBY'S

Sotheby Parke Bernet is an auction house with headquarters in London and branches in many of the major cities of the world, including New York City, where the action I want to describe now takes place. Sotheby's serves the useful function of selling for others at auction such exotica as

Chinese porcelains, vintage French wines, classical art, and oriental carpets, which I shall use in the discussion to follow. Naturally, one thinks of auctions as a place for such a find. At Sotheby's this is true only now and then.

If you inspect one of Sotheby's auction catalogs for oriental carpets you will find entries that read something like this.

■ 121 ANTIQUE KAZAK CLOUDBAND RUG

The three octagons in blue against a shaded red ground contain typical cloudband ornaments in red and ivory. All enclosed in a leaf and wineglass border. Approx. 4 ft 3 in.×6 ft 5 in. (130×196 cm).

If you look up item 121 in the back of the catalog you will find that Sotheby's predicts that the rug will bring $2,500–3,000. Apparently a bargain will be something less than the lower bound of this estimated range; the question is how much less? In working toward an answer to this question, it is important to know certain things about the catalog listings and how Sotheby's auctions operate. The symbol ■ before the item number means that the item carries a "reserve," a confidential price below which it will not be sold. In recent times, at least, every item in the rug catalogs has been marked with this symbol. If there were a way to find out the secret reserve price, a bargain would be a purchase at or just above that value.

Parameter Estimation

Getting technical for just a second, the problem of finding out the value of the reserve is another illustration of the inferential function of statistics that I mentioned at the end of Chapter 1, the procedure of *parameter estimation*. As a reminder, a *parameter* is the value of some measure for some *population*. In this case, the parameter will be the average reserve price, and the population is all the rugs sold at Sotheby's New York auctions of oriental carpets.

The parameter in question (the reserve) has to be estimated because of the way the auctions run. Sotheby's bid on items themselves until the bidding passes the reserve price. If the bidding does not pass the reserve, Sotheby's will have made the high bid and "bought" the item. This means that an item which appears to go for a low price may have been sold at that price or it may not have been sold at all, the auction house having made the highest bid. When this happens, the fact is not announced.

If one goes to a few auctions and notes the prices at which each article sells (or seems to sell), this information will put one in a position later to estimate an average reserve price. Following each auction, Sotheby's distributes lists of the selling prices of all items actually sold. The lists exclude the items not sold because the "selling price" was less than the reserve. Such items do not appear on the list of prices. If one has been to the auctions, however, and noted the prices, the necessary data are available. I now turn to an analysis of such information that I collected myself. The data came from four of Sotheby's New York auctions of oriental rugs and carpets that I attended over the span of about a year's time. I bought a couple of things and did not realize that I had been doing research for this book until after the data had been collected.

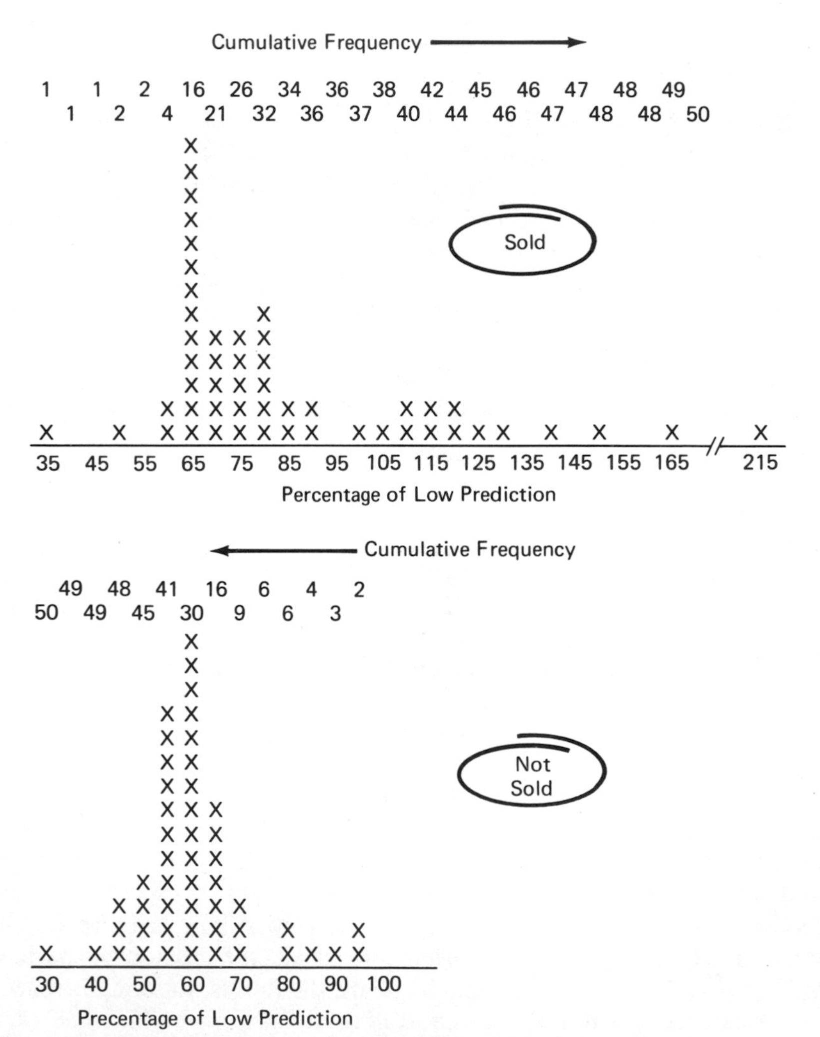

Figure 3–7 Frequency distributions which show "selling price" of 50 oriental rugs that were sold and 50 that were not sold at Sotheby's as a percentage of the lower limit of an estimated price range. The cumulative frequencies at the top of the two figures simply add the frequencies from left to right (above) and right to left (below).

At each auction, as almost everyone attending does, I jotted down the selling price of each article. These data provided the sample from which this parameter was estimated. When the actual price lists became available, I proceeded as follows: first I located 50 rugs that seemed at the time of the auction to have been sold but had not been (they were omitted from the price list) and noted their "selling" prices, which I expressed as a percentage (rounded to the nearest 5%) of the lower limit of the predicted range. Then I located 50 other rugs that actually had been sold, whose predicted range of selling prices were individually about (almost always exactly) the same as the rugs that had not been sold. Again I expressed their selling prices as percentages of the lower prediction. This procedure of matching rug for rug in terms of predicted selling price was to control for the effect of that variable upon the data.

The frequency distributions of these two sets of percentages appear in Figure 3–7. For the moment please ignore the "cumulative frequency" data in the figure. I will get back to them. Considering the two sets of data together, it is obvious at once that the reserve price must vary from item to item. The two rugs that sold for 50% of the lower predicted price or less must have had very low reserves. The six rugs that did not sell for 80% or more of the lower bound may have had fairly high reserves, or they may have been rugs that were not sold for some other reason. For example, the buyer may have failed to pay for them. As regards an average reserve, the data strongly suggest that it is something like 65% of the lower bound. Thirty-four of the 50 rugs not sold had high bids of 60% of the lower prediction or less. Forty-one of the 50 rugs that were sold had bids of 65% of this value or more. The sharp drop in the 60–65% "not sold" distribution and the sharp rise from 60–65% in the "sold" distribution confirm this impression.

Cumulative Frequency Distributions

Another way of presenting these data makes the same point even more clearly. This involves the cumulative frequencies entered in Figure 3–7. Inspection of these data will show that I have simply added the frequencies from low percentages to high ones for the "sold" distribution. This shows the increasing numbers of rugs sold with higher and higher bids. For the "not sold" distribution, I added frequencies in the other direction to show the increasing numbers of rugs not sold with lower and lower bids. Figure 3–8 presents these data in graphic form, technically as *cumulative frequency distributions*. The point of interest is at the intersection of the two functions, where the "not sold" curve cuts the "sold" curve. The value on the abscissa corresponding to this point is an estimate of the reserve price. As before, it turns out to be about 65% of the lower bound of predicted range of selling prices.

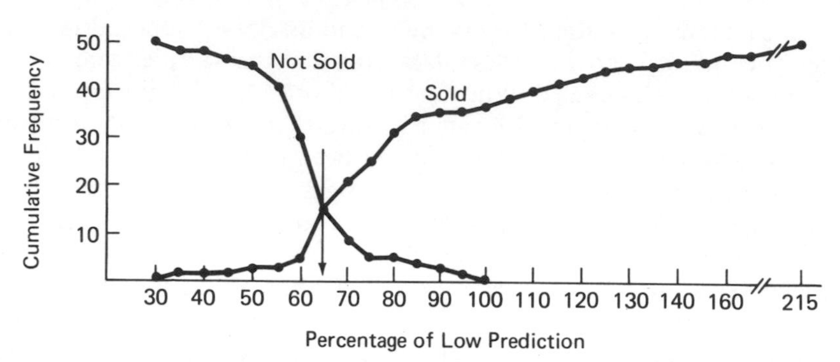

Figure 3–8 Cumulative frequency distributions. The numbers plotted appear above the two graphs in Figure 3–7.

A Word of Caution

The data I have just presented are real and I would bet that my estimated average reserve price is about right. Before you rush to Sotheby's next auction of items you covet, expecting to find a bargain, however, there are several considerations to put before you. The first is that my calculations may be way off for items other than oriental rugs, and they may change with time. A second fact is that reserves vary. All that I have been able to do is to estimate a rough average value. I have looked at the items that apparently had high reserves. Without exception they were standard examples of desirable types of rugs. Finally, current fashion pushes the prices of certain classes of items up well beyond the top of the estimated price range.[6]

SUMMARY–GLOSSARY

Frequency distributions provide a graphic display of the variations in some characteristic that exist in a sample or population. Although the distribution of major interest to statistics is the normal distribution, other forms of distribution occur. This chapter has presented examples of some of them.

[6]For those who know about rugs. Some of the items that had high reserves were a very impressive old Kashan, a Karachopf Kazak, a silk Heriz, and a new Nain. Those that seemed to be bringing prices above the top of predicted price range were: (1) any Caucasian rug in good condition; (2) all the Sennehs and Feraghans with the central diamond-shaped medallion and the Herati-covered ground (the item in Figure 3–7 that went for 215% of the low prediction was a Senneh of this type on a colored silk warp; the prediction was $3,000–$3,500, but it brought $6,500); and (3) in recent auctions, any large old Sarouk in good shape.

The important concepts in the chapter are defined below. Where concepts from previous chapters have reappeared in this one, I have defined them again.

Frequency distribution. A graph showing the distribution of some characteristic. A measure of the characteristic appears on the X axis of the graph. The number of cases, percentage of cases, or proportion of cases (generally, frequency) appears on the Y axis.

Frequency polygon. A frequency distribution formed by connecting the points in the graph with straight lines. This creates a many (poly)-sided (gon) plane figure. See Figure 3–1. Contrast with *bar graph.*

Bar graph. A form of frequency distribution in which rectangles (bars) are used to represent numbers of cases. See Figure 3–1.

Histogram. A bar graph.

Class interval. A range of scores lumped together for purposes of making a frequency distribution. Scores may be grouped by twos, threes, or any convenient interval size.

Unimodal distribution. A frequency distribution with one mode.

Mode. Literally, the most frequent score in a collection of scores. By convention, however, a set of scores is often said to have more than one mode, if there are two or more concentrations of scores.

Bimodal distribution. A frequency distribution with two modes. There can also be distributions with several modes, in which case they are referred to as *multimodal.*

Symmetrical distribution. A distribution where the shape on one side of the mode is the mirror image of the shape on the other side.

Skewed distribution. A distribution that is not symmetrical, having a longer "tail" on one side than the other.

Positively skewed distribution. A skewed distribution where the longer tail points in the direction of the high values of X.

Negatively skewed distribution. A skewed distribution where the longer tail points in the direction of the smaller scores.

J curve. Often encountered in studies of conformity, a sharply skewed distribution where the mode corresponds to adherence to a norm and increasingly nonconforming behavior occurs with decreasing frequency.

Hypothesis testing. The evaluation of hypotheses about a population on the basis of statistics obtained on samples.

Null hypothesis. The hypothesis of "no differences" often tested in the analysis of empirical data.

Level of Confidence. The confidence with which the null hypothesis can be rejected on the basis of a statistical test, the probability that an obtained difference would occur by chance if the null hypothesis were true.

Parameter estimation. The estimate of a value for the population (parameter) on the basis of a measure (statistic) for a sample.

Cumulative frequency distribution. A frequency distribution in which the abscissa is a score or measure, as is true of other frequency distributions, but the ordinate is cumulative frequency. Cumulative frequency is obtained by successively adding the frequencies of scores from low to high or high to low. See Figure 3–7 for an illustration of cumulative frequency and Figure 3–8 for an example of a cumulative frequency distribution.

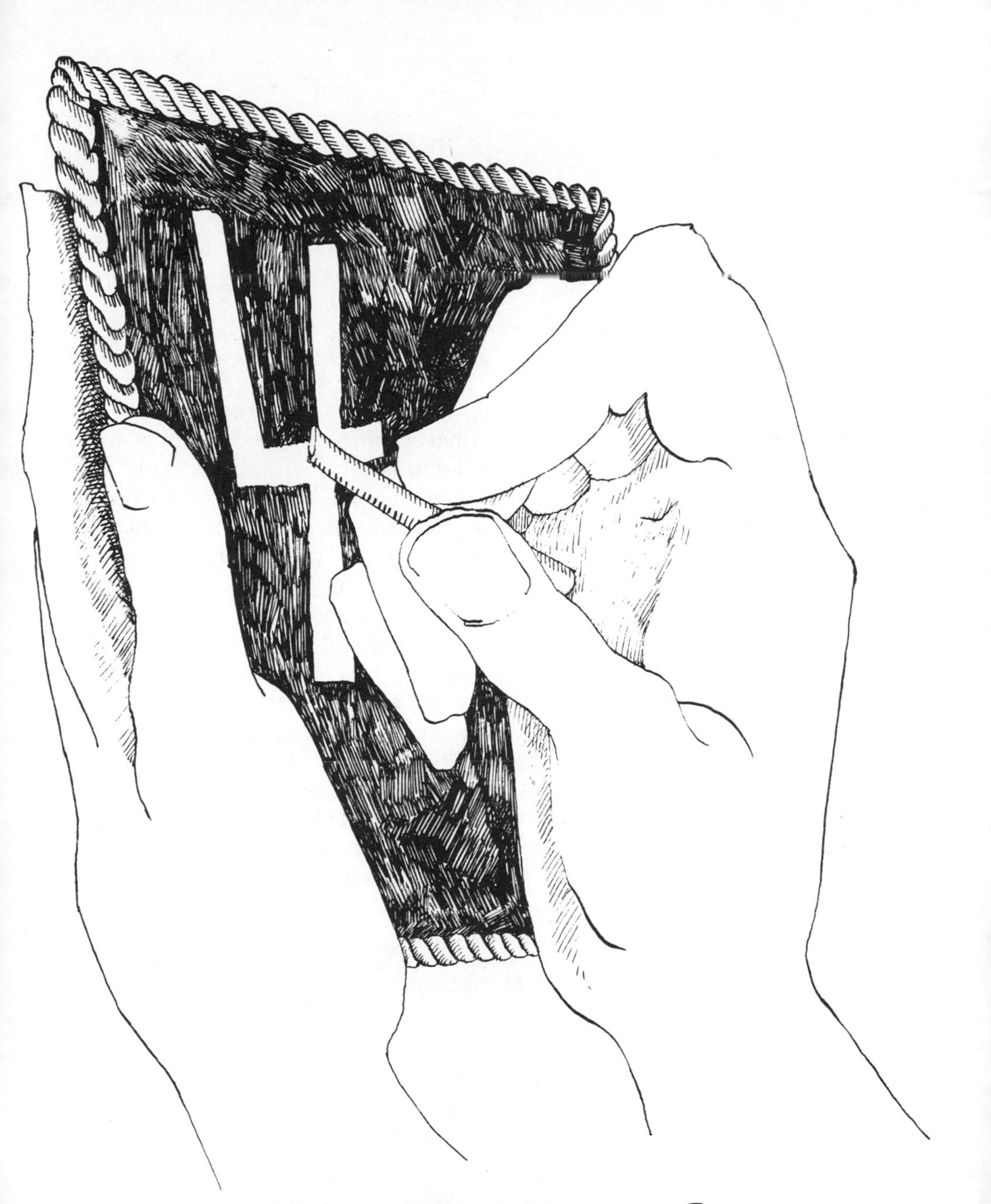

The Tasks of Science

Back in the thirteenth century King Frederick II of Germany "wanted to find out what kind of speech and what manner of speech children would have when they grew up if they spoke to no one beforehand. So he bade foster mothers and nurses to suckle their children, to bathe and wash them, but in no way to prattle with them, for he wanted to learn whether they would speak the Hebrew language, which was the oldest, or Greek, or Latin, or Arabic, or perhaps the language of their parents, of whom they had been born. But they laboured in vain because the children all died. For they could not live without the petting and joyful faces and loving words of their foster mothers."[1]

Seven hundred years later it seems unlikely that the children died because no one spoke to them, although death from more general psychological neglect is a recognized possibility. For obvious ethical reasons such statements are not based upon experiments, but upon less formal observations. For example, deaf children are not "prattled with" in any real sense, but they survive. But children who do not experience "petting and joyful faces" sometimes develop serious physical disorders. Such observations lie behind the plan of an imaginary study that I will describe now.

THREE-STAGE EXPERIMENTAL DESIGN

As I mentioned above, there have been no true experiments on the topic under discussion. But suppose there had been. What would such investigations have been like? The simplest possible studies would have followed a standard three-stage experimental design.

Stage I: Equating Groups

The study would begin with the selection of two groups of babies destined to be treated differently in the second phase of the experiment. In

[1] L. J. Stone and J. Church, *Childhood and Adolescence* (New York: Random House, 1973), p. 104.

this first stage the most important consideration is that the two groups be similar in terms of the effects of interest. If they were not roughly equal, any "results" of the experiment might reflect these initial differences rather than any effect of the experimental treatment.

A particularly dramatic example of such an error comes from a medical experiment on diptheria vaccine carried out many years ago. Several hundred patients admitted to a hospital with diptheria were in an *experimental group* who received the vaccine. A *control group* did not receive it and had only normal hospital care. The results were discouraging: 16% of the patients receiving the vaccine died as compared with 8% who did not receive it. These results could have been taken to mean that the vaccine killed patients instead of curing them. The true explanation, however, was that the hospital had been giving the vaccine to the patients who were most seriously ill and putting the milder cases in the control group. With this initial difference between the groups *any* conclusion about the general effectiveness of the vaccine would have been unjustified.

Stage II: Experimental Treatment

Experiments of the type I am describing in this section are sometimes called *bivalent* (*two-value*) *experiments* because they involve just two values of some independent variable, typically represented by two different treatments of an experimental group and a control group. Pursuing our King Frederick example, these two experimental treatments might go as follows. The infants in the *control group* receive "normal mothering," in which the foster mothers do all the things mothers do with their babies—talk to them, sing to them, fondle and rock them, and of course take care of their biological needs. The infants in the *experimental group* receive only the care required to satisfy these biological needs.

These treatments would go on for a specific amount of time decided at the beginning of the study. This feature of the experimental plan involves a risk. The time selected might be too long or too short to obtain an indication of the effect of the different procedures. The alternative, of running the experiment until an effect occurs or until the experimenter is convinced that nothing will happen, might work in this imaginary study, as the results to be described will indicate. Too often, however, such a procedure involves the risk of stopping the experiment at the point when the results are what the experimenter wants them to be.

Stage III: Evaluation

Bivalent experiments conclude with a comparison of the two groups. In the example being presented, the comparison might be in terms of qualitative statements something like this. *Control group:* "Normal, happy babies, appetites good, level of activity high, interested in the world around them, no signs of physical or mental disorder." *Experimental group:* "Babies vary greatly. Some extremely tense and given to wild screaming and vomiting. Others lethargic, refuse to nurse, fall into stuporous sleep, and seem near death."

These imaginary results for the experimental group describe the symptoms of *marasmus*, a condition which 75 or 100 years ago accounted for 50% of the infant mortality rate. It is now reasonably well established that psychological neglect is a contributor to, if not the main cause of, marasmus.

Summary

The plan of bivalent experiments can be summarized in neat diagrammatic form, as in Table 4–1. The cells of the diagram can be filled in with examples from the study of parental neglect in the following way. *Stage I* involved the selection of two groups of babies who were comparable in terms of health. In *stage II* the babies in the control group received normal mothering; those in the experimental group received only essential physical care. One thing which this description reveals is that the designations "experimental" and "control" can be pretty arbitrary. They could easily be reversed in this example. Observations in *stage III* showed the powerful effects of the differential treatment.

TABLE 4–1
Plan of a Bivalent Experiment

Experimental group	Measurement to equate groups	Experimental treatment	Measurement to assess outcome
Control group	Measurement to equate groups	Control treatment	Measurement to assess outcome

Confounding

The logic of interpretation for the results of any experiment in the life sciences is straightforward: If the individuals in the experiment were the same at the beginning (stage I) but different later (stage III), the difference must be the result of the differences in experimental treatment (stage II). This argument is sound, however, only with the important proviso that some unintended treatment did not accidently produce the outcome. In our example if the babies in the experimental group had received inadequate nourishment as well as lack of mothering, this could account for the results.

The wonderful world of advertising is full of examples where this proviso fails to be met. Recently there has been a TV ad for a particular shave cream in which we are offered the following "experiment" urging us to buy the advertiser's product. The subject of the experiment is a man with visible need for a shave. Both sides of the subject's face are prepared for shaving, one side with the advertised product, the other side with an unidentified "other" preparation. Then both sides are shaved, and the comment points out that they bear down particularly hard on the side with the product being advertised. Presumably the point of this is to suggest that shaving with the advertised brand is a comfortable experience. Then comes the test. The announcer scrapes the edge of a credit card against the grain of the subject's beard on both cheeks. The side shaved with the product being advertised sounds smoother. Conclusion: The advertised shave cream leads to a better shave.

The clearest error of experimental design in this ad is that bearing down particularly hard during shaving may have created the smoother shave; the shave cream may have had nothing to do with it—even assuming that no other sources of error (e.g., the placement of the microphone) are responsible. Technically, we call this an error of *confounding*. More than just the intended experimental treatment (the brand of shaving cream) has been introduced in stage II of the experiment. The independent variable (differences in brands of shaving cream) has been *confounded* with another variable (force applied to the razor) that might have produced the effect.

Another example involves people's preferences for soft drinks. In what it described as a blind taste test, Pepsi Cola has presented subjects (all Coca-Cola drinkers) with two colas to compare. A glass marked "M" contains Pepsi Cola; a glass marked "Q" contains coke. The subjects are asked to state a preference. Results: "Half the Coca-Cola drinkers in the Dallas–Ft. Worth area actually prefer Pepsi." There is more than one thing wrong with this experimental test and the conclusion to which it leads.

For one thing, the results may mean that people cannot discriminate between Pepsi and Coke. Since they have to choose one glass or the other, however, half of them pick the glass containing Pepsi and thus "prefer" it. Or the experimental procedure may contain an element of confounding, as the Coca-Cola people claim in their counterattack: "Here's a fascinating report. Two glasses, one marked M and the other marked Q. Both glasses contain the same thing. Coca-Cola. We asked people to pick the one that tasted better. Most of them picked M even though the drinks were the same. You know what that proves? It proves that people will pick M more often than Q. So M has an advantage."

MULTIVALENT EXPERIMENTS

The trouble with bivalent experiments is that the *most* they can tell you is that some variable either does or does not make a difference. For example, the imaginary experiment on parental neglect indicates only that mothering (incidentally even if such attention happens to be provided by a father) is important for a child's well-being. But the interpretation depends upon the details of results. In bivalent experiments this fact poses a special problem.

Suppose (if necessary) that you are a parent. Do the data on marasmus mean that you have to spend every waking moment providing your baby with Tender Loving Care? What about the common belief that too much TLC is a bad thing—that babies can be "spoiled" by attention? These two sets of ideas together suggest that the relationship between the adjustment of the child and the amount of parental attention may be nonmonotonic. Up to a point babies may benefit from parental attention; with more attention they may suffer by being spoiled. Perhaps the relationship is something like that shown in Figure 4–1.

Suppose that it is. Think of what this means for the possible outcomes of bivalent experiments comparing the effects of two different amounts of parental attention. As I have shown in Figure 4–1, they could be anything. Depending upon the values studied, increased attention could have a beneficial effect, no effect at all, or a harmful one. The general point to draw from this is that it is often important to study the influences of

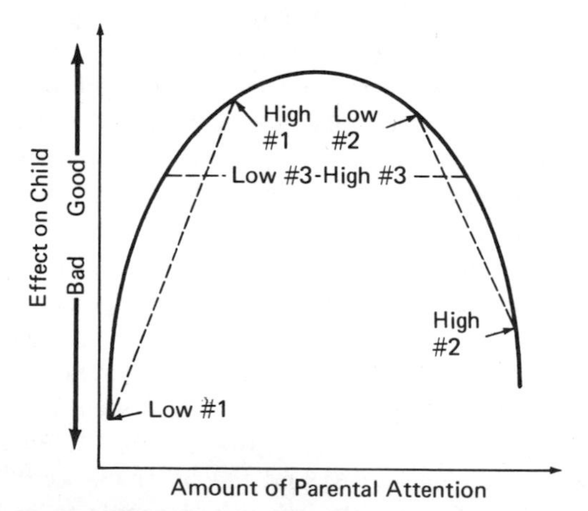

Figure 4–1 The trouble with bivalent experiments. If the true relationship between some independent and some dependent variable is nonmonotonic, the general trend implied by the results of a bivalent experiment will depend upon the two values selected. Low 1 and High 1, Low 2 and High 2, and Low 3 and High 3 are three pairs of such points in imaginary experiments. As the placement of these points indicates, the results would imply increasing, decreasing, and no effect, respectively.

several values of an independent variable in order to get a better picture of the relationship involved. Such experiments are called *multivalent* (many-value) experiments.

The general strategy employed in multivalent experiments is straightforward and obvious: the experimenter tries out different values of an *independent variable* and observes the effect on a *dependent variable.* The examples I have chosen will make the procedure more concrete. They will also serve to illustrate two important procedural distinctions between (1) *manipulated* and *selected independent variables* and (2) *between-subject* and *within-subject designs.*

Pupillary Dilation and the Mind of Man

It has been known for centuries that the size of a person's pupils reveals something about his frame of mind. For example, Chinese sellers of jade are said to watch the pupils of a buyer closely, to use pupillary dilation as a sign of interest in a particular stone and to set the price accordingly. The pupils of hungry people also dilate to the sight of food, and this fact leads us to our first example of a multivalent experiment. In this experiment the investigator presented colored slides of different foods (spaghetti, cold cuts, steak and potatoes, roast beef, and roast turkey) to 20 people who had gone without eating for 5 to 8 hours and measured the change in pupillary diameter in millimeters. The subjects' pupils dilated to all of these pictures of food but the amount of dilation was different for different foods.[2]

Before I present the results in more detail it may be a good idea to use this example to review some important points: (1) the independent variable in this experiment is the kind of picture presented (spaghetti, cold cuts, etc.); (2) the dependent variable is change in pupillary diameter in millimeters; and (3) the study is an example of a within-subjects experiment. There were 20 participants and all of them saw all five slides. To do the same experiment between subjects there would have to be five groups of subjects, each group seeing just one slide.

The obvious advantage of within-subjects procedures is their efficiency. A potential disadvantage and sometimes a feature that disqualifies the within-subjects design is that exposure to one value of the independent variable may influence reactions to the other values. If one wanted to study the relative values of two methods of learning the content of a paragraph—say, rote memorization versus reading for ideas in fourth-grade pupils—the experiment would have to be done with a between-subjects design. If there were just one paragraph to learn, the pupils could not

[2]E. H. Hess, The role of pupil size in communication, *Scientific American*, **232**, 1975, pp. 110–118.

master it both ways. If there were several paragraphs, it would be very risky to ask for one method of study on some paragraphs and the other method on others. If the method of study makes a difference, the pupils would find that out and almost surely use the more effective method, no matter what the experimenter told them to do.

Now from pupils and paragraphs back to pupils and palatability. The results of the experiment were as shown in Table 4–2, with the different foods listed in order of the amount of pupillary change they produced and presumably, therefore, in order of tastiness. Obviously, the different foods (independent variable) produced different amounts of pupillary dilation (dependent variable).

TABLE 4–2
Palatability and Pupil Size

Food	Pupillary Increase (mm)
1. Cold cuts	.08
2. Spaghetti	.12
3. Roast beef	.21
4. Roast turkey	.26
5. Steak and potatoes	.39

These results will make it possible to call attention to a major problem in the interpretation of data—the problem of generality. There is reason to suspect that the specific ordering of foods from least palatable (cold cuts) to most palatable (steak and potatoes) suggested by this experiment has very limited generality. For one thing, within-subject experiments create this problem: since the same individuals were in every experimental condition, it follows that the obtained order may apply only to these subjects and, even worse, only when these subjects are tested with these specific pictures. Different photographs might have produced very different results. To anticipate a very important point of interpretation, the results of experiments can be generalized only to populations of which the materials and subjects involved are a representative sample.

Happy Eyes

The preceding experiment was done within subjects. For an example of a between-subjects experiment, let us consider another experiment on pupil size, this time on the relationship between pupil size and the judgment of whether a face is a happy face. There is a good bit of informal evidence that faces with large pupils seem happier than faces with small pupils. For example, students asked to draw in the pupils on faces like those in Figure

Figure 4–2 From "The role of pupil size in communication," by Eckhard H. Hess.

4–2 will make them larger in the smiling face than in the frowning one. For another example, happy faces in children's books are apt to have larger pupils than unhappy or angry faces. With this as background the investigator set out to study the development of this tendency. He showed nine groups of individuals, ranging from 6 years of age to 22, drawings that were the same except that the pupils were of different sizes. His data were the percentage of people in each group who judged the drawing with the larger pupils to be happier.

Again, before presenting the results of the study, it will be useful to point out the important features of the experiment: (1) the dependent variable in the experiment is the percentage of people in each group selecting the drawing with the larger eyes as happier; (2) the independent variable is the age of the subjects in each group: 6, 8, 14, etc.; and (3) in this case the independent variable is a *selected variable*. Individuals having the desired value of the independent variable were chosen for study. This

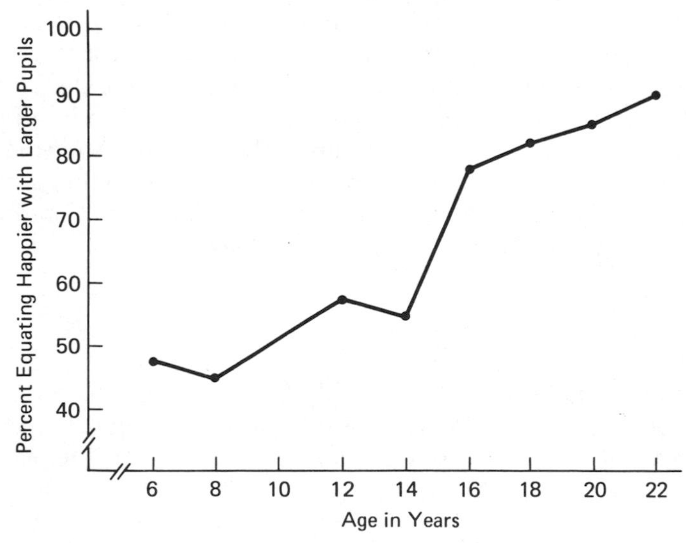

Figure 4–3 Judgment of happiness on the basis of large pupils increases with age.

is to be contrasted with the photographs of different foods in the previous study. In that case the independent variable was a *manipulated variable*. The experimenter determined these values.

There is rarely an option in the matter of deciding whether to use a selected independent variable or a manipulated one in an experiment. The selected variables are almost always aspects of the personal makeup of people, over which the experimenter has no control—age, sex, race, religious beliefs, amount of education, income level, intelligence, and the like. Manipulated independent variables are features of situations that the experimenter can control.

The results of the experiment appear in Figure 4–3 (p. 69), where it seems clear that, with minor irregularities, the function relating percentage of subjects judging the face with the larger pupils as "happier" increases steadily with age.[3]

FACTORIAL DESIGNS

Multivalent experiments provide a picture of the relationship between some dependent variable and some independent variable. It is important to realize, however, that the nature of that relationship will depend upon the details of the conditions under which it was obtained. A common finding is that the relationships do not hold up well if one alters other aspects of the experimental situation.

Another way to put it is that most phenomena are complexly determined. This statement means two things: (1) the phenomenon has numerous causes, and (2) these causes interact. The first of these points is probably obvious. The second may not be, but an understanding of the concept of *interaction* is essential to the appreciation of the idea of complex causality. Because this concept is so important, I shall describe it in great detail. You may come away from your reading of this section in the same frame of mind as the little girl who began her book report: "This book told me more about butterflies than I wanted to know."

Main Effects and Interactions

The points I have made so far imply that the studies to be considered now, *factorial designs*, will have more than one independent variable. Such studies make it possible to ask questions of two sorts: (1) Does each of the variables have an individual effect? (2) Does the magnitude of this effect

[3]Ibid.

depend upon the values of the other variables. Technically the first of these questions is about *main effects*; the second is about *interactions*.

In order to make these ideas more concrete I will present some examples, together with simplified results, of actual experiments—simplified in order to bring the concepts of main effect and interaction into sharp focus.

Learning to Read. Possibly the most serious educational problem of the world today involves the development of better ways to teach people to read. Over the years there have been many proposals in the area, and many fads, but little in the way of real accomplishment. Studies such as the following identify a part of the reason for our lack of progress.

This study compared the effectiveness of two methods of teaching reading. One was the phonic method, in which children sound out the words they try to read. The other method was the sight method, in which children try to get a general visual impression of the word and to learn to read that way. Notice that the method of teaching reading is a manipulated independent variable.

A second variable in the experiment was a selected one. Half of the children were very gifted in general intelligence. The others were nongifted. Half of the gifted group and half of the nongifted group were taught by the phonic method; the other half of each group were taught by the sight method.

Technically this experiment is a *2 by 2* (or 2×2) *factorial design*. Two values of one independent variable (method of teaching) are combined with two values of a second variable (ability level). The design of this particular experiment, together with a set of idealized results, appears in Table 4–3.

You should notice three things about these data:

1. Averaged across teaching method there is a substantial difference between the gifted and nongifted groups in speed of reading—55 versus 20 words per minute. This difference reveals a *main effect* of ability level.
2. There is also a main effect of teaching method averaged across ability levels —45 versus 30 words per minute.

TABLE 4–3
Average Reading Speed (words per minute)

	Method of Teaching		
Ability level	*Phonic*	*Sight*	*Average*
Gifted	60	50	55
Nongifted	30	10	20
Average	45	30	

3. The overall difference produced by teaching methods differs for the gifted and nongifted groups. For the gifted children the two methods made less of a difference than they did for the nongifted children, who learned very much better with the phonic method.

This is to say that there is an *interaction* between ability level and teaching method. The effects of the different methods differ for the different groups. In terms of a table such as Table 4–3, the differences between columns differ from row to row: $60-50=10$ versus $30-10=20$; or the differences between rows differ from column to column: $60-30=30$ versus $50-10=40$.

Briefly Summing Up. A main effect of an independent variable exists when the different values of that variable produce different results when the data are collapsed (added or averaged) across values of the other independent variables in the study. In the reading example, the phonic method of teaching is superior to the sight method when the data are averaged for gifted and nongifted children. There is a main effect of teaching method.

Similarly, there is a main effect of ability level, because the gifted are superior to the nongifted when their reading speeds are averaged for the two different teaching methods.

An interaction exists when the effect of one independent variable depends upon the value of the other. There are always two ways to show an interaction in a 2×2 factorial experiment. To illustrate the first, there is only a 10-word/minute superiority of the phonic method over the sight method for gifted children, whereas for nongifted children, this difference is 20 words per minute. In short, the difference between the phonic and sight methods differs for different levels of ability.

Moving on to the second way of showing the existence of an interaction, there is a 30-word/minute difference between gifted and nongifted children when they are taught by the phonic method. When they are taught by the sight method, this difference increases to 40 words per minute. In short, again, the difference in reading between gifted and nongifted children differs for different methods of teaching.

In my experience the concept of interaction is difficult for some people. It is so important, however, that I shall persist in trying to make it clear. My next example, in which there are no main effects, only an interaction, may be of help.

Race and Mortality. In European cities whites outlive blacks largely because of the Caucasians' greater resistance to tuberculosis. In African cities blacks outlive whites largely because of the blacks' greater resistance to yellow fever. Suppose that the actual data are as shown in Table 4–4. In this table you should notice three things:

TABLE 4–4
Deaths per 100,000 People per Year

Race	Location		
	Europe	*Africa*	*Average*
Black	1,050	750	900
White	750	1,050	900
Average	900	900	

1. There is no main effect of race: death rates for blacks and whites, averaged for European and African communities, are identical.
2. There is no main effect of location: European and African death rates averaged for blacks and whites are also identical.
3. There is a substantial interaction, however: in Europe the death rate is 300/100,000 greater for blacks than whites, but in Africa, this difference is reversed. Or, looking at it the other possible way, for blacks the death rate is 300/100,000 greater in Europe than in Africa; for whites, this difference is reversed.

Pursuing the effort to make these materials clear, I turn now to an example that is almost the opposite of this one, to a case where there are two main effects and no interaction.

Work, Rest, and Productivity. There is considerable evidence to prove that the amount accomplished in a fixed amount of time by a person performing a physical skill improves if (1) the duration of work periods is short (the fixed amount of time is broken up by rests) or (2) the rest pauses between these periods of work are long. Suppose that the management of a factory producing electronic equipment studies the number of articles that its workers assemble by hand. The experiment is a 2×2 factorial involving long and short work periods separated by long and short periods of rest. After some amount of time involving the same total amounts of working time, the results are as shown in Table 4–5. In this case you should notice three things again:

TABLE 4–5
Number of Items Assembled

Length of Work Period	Length of Rest Period		
	Short	*Long*	*Average*
Short	120	150	135
Long	80	110	95
Average	100	130	

1. There is a main effect of the length of the work period. Averaged across short and long rest periods, the subjects *working* for short periods produced 135 items; those *working* for long periods produced only 95.
2. There is a main effect of the length of the *rest* period. Averaged across short and long work periods, subjects given short *rest* periods produced only 100 items; those given long *rest* periods produced 130.
3. There is *no interaction* between length of rest period and length of work period. Showing this two ways: (a) Increasing the *rest period* from short to long increases production by 30 items whether the work period is short or long; (b) increasing the *work period* from short to long decreases production by 40 items whether the rest period is short or long.

Overlearning and Memory. Our final example is again one for which there is substantial evidence for the relationship described. Suppose that two groups of students learn the same set of materials for a course examination. One group studies until they have learned the materials completely, but the other group continues to study, spending, let us say, as much more time as it took them to master the materials in the first place. Let us call the first group a *learning* group, the second an *overlearning* group. Suppose now that the two groups take tests covering this subject matter the next day and then a month later. If scores are percentage of materials correct, the results might be like those shown in Table 4–6. As with the first example presented there is evidence for three influences in these data: (1) a main effect of amount of study, (2) a main effect of time between study and test, and (3) an interaction.

TABLE 4–6
Percentage Correct on Tests

	Time Between Study and Test		
Group	1 *Day*	30 *Days*	*Average*
Learning	90	50	70
Overlearning	95	85	90
Average	92.5	67.5	

Higher-Order Factorial Designs

In order to keep the presentation of factorial designs to fundamentals I have limited the discussion so far to 2×2 (read "two by two") factorials. It is important to know, however, that factorial designs are usually more elaborate than this. Usually, there are more than two values of the two independent variables, or more than two independent variables, or both. The 2×2 factorial design has two rows and two columns as I have presented these designs. In effect, a 2×2 factorial crosses one bivalent experiment with another. The number of rows and columns can be as large as necessary, however. In effect, such a more elaborate design crosses one multivalent experiment with another. The result is an experiment that requires the number of groups needed to perform one of the two multivalent experiments multiplied by the number of groups required to do the other. These comments refer to an experiment with only two variables. Adding additional variables (with some number of values of each) carries the multiplication further. An experiment with three variables would require the following number of groups: the number of values of variable 1×the number of values of variable 2×the number of values of variable 3. Still higher numbers of variables can be manipulated, but, as you can see, they are rapidly going to get cumbersome to describe and interpret. A 4×4 factorial is as complex as I intend to get in this book. Such an experiment is described in the computational appendix.

A Different Kind of Summary

The four hypothetical studies presented in the previous section were selected to give realistic examples of a few of the combinations of main effects and interactions that can emerge from an experiment. In presenting these examples I put the data into tables. Another way to present the data would be in the form of graphs, which provide a different way of summarizing the meanings of the concepts of main effect and interaction. Figure 4–4 presents the data just discussed in such a form.

A graph showing the results of a factorial experiment with two independent variables is constructed as follows. The horizontal axis of the graph represents one of the two independent variables. The vertical axis represents the dependent variable. The second independent variable is dealt with by presenting two functions in the graph, one for one value of this second independent variable, the other for the other value of the second independent variable. The fact that there are two independent variables in these factorial experiments means that there are two ways to draw these graphs, depending upon which independent variable appears on the abscissa of the graph. Figure 4–4 presents the data both ways.

In this way of presenting the results, a main effect appears in one of two ways. The first of these ways is as a vertical separation of the two curves in the body of the graph. Since the data are plotted with the two different independent variables on the base line in the separate graphs, it is easy to see that both independent variables are producing a main effect in all the graphs except the two for death rates in blacks and whites in Europe and Africa.

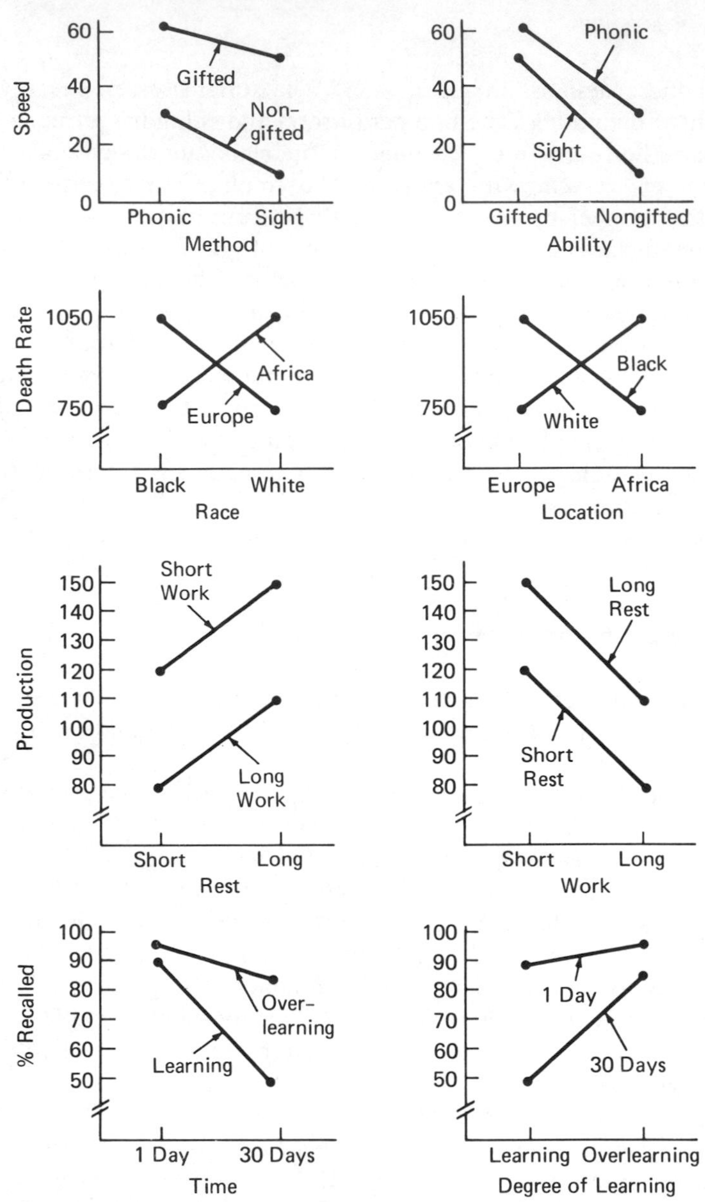

Figure 4–4 Graphic representations of outcomes of 2×2 factorial studies. The general method for making such a graph is to let the Y axis represent the dependent variable in the study and the X axis one or the other of the two independent variables. The second independent variable appears as a designation for the two functions in the figure. This figure plots the data discussed earlier both ways. Main effects appear in two ways: (a) as a general trend upward or downward if you imagine the two functions averaged (all graphs except for the two for race and death rate) or (b) as a separation of the points that would be produced if you averaged the two functions. An interaction takes the form of curves that are not parallel (all except the curves for work and rest).

The second way in which a main effect reveals itself is by a tendency for the two curves when averaged together to show a trend one way or the other. An inspection of either of the two figures for each set of data will lead you to the same conclusion as the two graphs for each set just have.

An interaction is present whenever the curves are not parallel. This means that the effects of the values of the first variable are different for different values of the second, and vice versa. Inspection of the graphs in Figure 4–4 will show that there are interactions in all the graphs except those for the effects of work and rest. I strongly urge you to study these materials until they are clear. A little persistence will see you through.

QUASI-EXPERIMENTAL DESIGNS

In New England there is an organization of young psychologists called the Psychological Round Table (PRT), to which I once "belonged." *Belonged* is an inappropriate term in a way. There is no formal membership, although people who are invited once to the meetings of the group tend to be invited again. The rules of the organization, such as they are, are simple. Plans for the meetings are made by a self-perpetuating group of six people, the "secret six" or "autocratic minority." Each spring this group invites 40 or so psychologists to attend a meeting and, if they wish, offer a "revelation," more prosaically a scientific paper. Invitations cease to come to an individual when he violates any of the three rules: failure to accept two invitations in a row, ceasing to do research, or reaching the age of 40. Such a person is "superannuated" and never hears from PRT again.

One of the highlights of the meetings is a banquet at which members receive irreverent "awards" for their accomplishments and at which there is delivered the William A. Hunt Memorial Lecture, named in honor of the first person to give one. The William A. Hunt Memorial Lecture is saved from pornography only by the caliber of intellect that gives rise to it.

Probably the author of the fragment of a William A. Hunt Memorial Lecture that I wish to use to introduce my topic should remain anonymous. His talk concerned a large-scale project he claimed to be working on. Heavily financed by the National Science Foundation, the object of the investigation was to acquire quantitative data on the dimensions of the American penis. One can hardly imagine the intensity of the scientific activity that went on in preparation for this project. There were deep problems of operational definition that had to be faced. In references to penile dimensions should the definition be in terms of length, volume, or perhaps some more highly derived form of measurement? Then, of course there were all the problems of constructing and calibrating experimental apparatus. Finally, these seminal issues were settled and the investigators were ready to run their first subject. He was chosen from the population with careful attention to random selection. His informed consent to participate was secured. He was invited to the laboratory and the study was about to begin. "But right there," in the words of the speaker, "we ran into trouble with the Heisenberg Principle: The operations of measurement were influencing the quantity we wanted to measure."[4]

Reactive Effects of Measurement

This contrived example anticipated by some 20 years a program of more serious work directed at an attempt to handle the fact that people in experiments on their behavior may react quite differently from the way they do under more natural circumstances. Experiments produce in people a sort of "guinea pig effect." They try to present themselves in a particularly favorable light or more rarely to present themselves unfavorably. In short, there are *reactive effects* of behavioral measurement which may distort the results of experiments.

Unobtrusive Measures

Attempts to get around such problems have led to improvements in studies that use the methods of naturalistic observation. Such *nonreactive research* employs what are called *unobtrusive measures*. These are records of behavior collected for purposes other than research. Often they are available to anyone. For example:

1. Interest in various advertisements in a magazine was studied in a library by counting the number of different fingerprints on the pages that carried the different ads.

2. A study in the United Kingdom showed that power blackouts tended to occur at the same times as the breaks in TV programs. Presumably people used the interruptions to plug in their tea pots or to go to the toilet,

[4]*A Word to the Wise.*[5] The speaker knew that this is not an accurate statement of the Heisenberg Principle, which is something more like this: It is impossible in terms of ordinary geometric conventions to assert the position and motion of a particle at the same time, for the more exactly one factor can be stated, the less exact is the statement of the other. The "operations of measurement" way of putting it is, however, a common metaphor.

[5]James Thurber once said that a word to the wise is not sufficient if it doesn't make any sense. I suspect that the actual statement of the Heisenberg Principle will prove the truth of Thurber's observation for many of you.

turning on the lights. The surge in power consumption was responsible for the power failures.

3. The method I used to estimate the reserve price at the auctions of Sotheby Parke Bernet used information that was a matter of public record (p. 54 f.).

4. At the Chicago Museum of Science and Industry investigators used the rate at which floor tiles wore out as an indication of interest in various exhibits. Those before a chick-hatching display lasted only about six weeks; those before other exhibits usually lasted for years.

5. Liquor consumption in a "dry" town was investigated by counting the bottles in people's garbage.

6. In a study of children's fear produced by a ghost story, the measure was the decreasing circumferences of the circle in which the children sat.

7. Once I did a small study to check on the common belief that changing answers on multiple-choice tests results in turning right answers into wrong ones by looking at answers involving erasures. Two-thirds of the changes were from wrong to right. The rest were divided about equally between right to wrong and wrong to wrong.

Such studies are sometimes referred to as *quasi-experimental designs. Quasi* means having some resemblance to the real thing. In these studies, for example, one might identify a "quasi" independent variable in most of them: the different ads, the time between TV programs versus other times, the various exhibits at the museum, and the passage of time during which the children were listening to the ghost story. The dependent variables in all cases are fairly straightforward indices of some aspect of behavior.

The Double-Blind Experiment

A somewhat different problem with experimental procedures is what has been called the *demand characteristics* of an experiment. I have already mentioned that subjects in experiments on testing tend to present themselves in the best light possible. Just what this best light turns out to be depends upon the individual's interpretation of what is expected of him, hence the expression "demand characteristics." To take just one important example, male participants in experiments involving electric shock often interpret the procedures as a test of their masculinity. For example, an experimenter might want merely to determine the speed with which a person can react to a shock that is unpleasant but not painful. As a preliminary this investigator must establish the level of such a shock. He presents a series of increasingly intense stimuli, asking the subject to tell him when the shock is so strong that he should cease. With his manhood apparently at stake, however, this hypothetical subject carries on beyond the point at which the objective requirements of the experiment would have allowed him to stop. The conditions of the experiment as he misinterpreted them seemed to demand it.

Another very different situation where demand characteristics are important is the medical one. The apparent demand placed on a patient under a doctor's care is to get well. Members of the medical profession recognize that these reactions can actually cause an improvement in physical health. Just as disease can be brought on by psychological factors, the same factors can lead to a cure.

In more objective terms what this means is that, when a group of patients take a drug and are helped by it, for some of them the help may be entirely psychological—technically a *placebo effect*. The word *placebo* is from the Latin and means "I shall please." The idea is that the patient tries to please the physician by getting better.

The situation in the doctor's office is complicated by the physician's own needs and expectations. The doctor needs to help his patients and uses a particular therapy because he expects it to do so. If he is an experimenter, the chances are that he is studying a particular drug because of the belief in its therapeutic value. He has what is more generally called an *experimenter bias*: the hope, belief, and expectation that a particular treatment will be effective.

The difficulties jointly produced by the placebo effect and experimenter bias have led to the creation of the *double-blind experiment*. Suppose that a medical researcher wishes to evaluate a drug. He gives some patients the real drug and others a placebo, a substance that looks and tastes like the drug being evaluated but without any expected curative power. The most common placebo is a sugar pill, which the patient believes is the real drug. The patients who participate in the study do not know whether they are in the drug group or the placebo group. They are "blind" as far as their group membership goes.

Beyond that, because of the problem of experimenter bias, it is important that the investigator also be blind with respect to which patients receive the drug and which receive the placebo. The easiest way to pull this off is apt to be for some third party to assign patients to experimental conditions and to provide the investigator only with a record of the outcomes. Studies employing the double-blind design have shown that both controls are important. New therapies introduced without such controls frequently claim a very high rate of cure. Adding the control for patients' knowledge reduces the effectiveness some; including the control for the experimenter's expectations tends to being the rate down still further.

SUMMARY–GLOSSARY

This chapter has covered some of the basic principles related to the design of experiments. An experiment is a device for making controlled observations upon the effects of variations in some independent variable. Experiments can vary greatly in degree of complexity. Some of the commonest types of experimental design are described in the following glossary of terms.

Three-stage experimental design. Implicitly at least, most experiments involve three stages: (1) a stage in which it is determined that the groups to be studied are initially equal with respect to the phenomenon under investigation; (2) a stage in which the independent variable takes on different values for different groups; and (3) a stage in which the effects of the variations in the second stage are assessed. If the experiment has been carefully done, differences in the third stage may be attributed to the differences in treatment in the second stage.

Bivalent experiment. An experiment with just two values of the independent variable. Often these experiments involve the administration of a special treatment to one group, and the withholding of that treatment from the other.

Control group. In an experiment, the group from which special treatment is withheld.

Experimental group. In an experiment, the group that receives special treatment.

Confounding. The accidental introduction of a factor that produces the effect attributed to the variations of the independent variable.

Multivalent experiment. An experiment with more than one value of the independent variable.

Independent variable. In discussions of experimental design, a condition that varies from group to group. Interest is in the effects of these variations.

Dependent variable. The phenomenon of interest in an experiment.

Between-subjects experiment. An experiment in which different subjects experience different values of the independent variable.

Within-subjects experiment. An experiment in which subjects experience more than one (conceivably all) values of the independent variable.

Selected independent variable. A condition that is varied by selecting subjects with the desired characteristics.

Manipulated independent variable. A variable whose value the experimenter can control and administer at will.

Factorially designed experiment. An experiment with more than one independent variable and two or more values of each. Each value of a given independent variable is combined with each value of every other independent variable.

2×2 *factorial design.* A factorially designed experiment with two independent variables and two values of each variable.

Main effect. The effect of variations in one variable in a factorial experiment, when these effects are collapsed (added, averaged) across all values of other independent variables with which the variable showing the main effect has been combined.

Interaction. The effect of variations in one independent variable changes with variations in the values of other independent variables in a factorially designed experiment.

Higher-order factorially designed experiment. A factorial experiment with more than two independent variables and more than one value of each variable.

Reactive effects of measurement. In research on human beings, a "guinea pig effect." Subjects do not behave normally or naturally when they know that they are being tested.

Nonreactive research. Research carried out in ways that avoid reactive effects of measurement and the demand characteristics of experiments. Usually this means research the subjects do not know they are participating in.

Unobtrusive measures. Measures obtained without the subjects' knowledge.

Quasi-experimental design. A study using the natural variations in an independent variable to determine the effects of these variations on a dependent variable.

Demand characteristics of an experiment. Related to reactive effects of measurement. The performance that human subjects believe to be expected (demanded) of them in an experiment.

Placebo. Most specifically, a pill that looks like a real pill but lacks the active agent of the real pill. More generally, any treatment that the subjects accept as an experimental treatment although it lacks the feature expected to have an effect.

Placebo effect. The influence of a placebo, which sometimes is as great as the influence of the experimental treatment.

Experimenter bias. The investigator's expected or hoped-for result of an experiment.

Double-blind experiment. An experiment in which subjects do not know whether they are in an experimental or control (placebo) group, to eliminate placebo effects, and the experimenter remains similarly "blind" to control for the effects of experimenter bias.

The Laws of Chance

To paraphrase Henri Poincaré, writing sometime around the turn of the century: How dare we speak of the *laws* of chance? Is not chance the opposite of lawfulness? Is not chance a word that stands for ignorance where lawfulness stands for what we know? Is it not, therefore, a contradiction to think of the laws of chance? The ancients did not have this problem because they distinguished between two kinds of phenomena, those which obey laws set down once and for all by the gods and those which are rebellious to all laws. These last phenomena they attributed to chance. Things were more sensible in those days. Law was law and chance was chance for all men and even for the gods.

But this is not our view today. We have become absolute determinists for most purposes. We may want to reserve the right of free will for the control of human conduct, but we let determinism rule unchallenged in the physical world. And this is why we have the contradiction. If every physical event is determined, what does it mean to say that red or black at the roulette table, heads or tails in a game of coin toss, or for that matter living to age 70 or 90 depends upon chance? First and last, what is chance?

FROM DETERMINISM TO PROBABILISM

Poincaré had pretty much the same answer to his question as we have today, but he spelled it out more fully than is fashionable now. All phenomena are determined, he said, but those that we call "chance" phenomena have two related characteristics. The first is that, although caused, "chance" phenomena have very small causes. The second is that these small causes act independently of one another.

All for Want of a Horseshoe Nail

Think of trying to balance a cone upon its tapered end. Release it and the cone topples immediately. Why is this? Because tiny imperfections, tiny imbalances, tiny vibrations destroy its equilibrium. If the cone were perfect and were acted on only by the pull of gravity, it would not fall.

Think of a perfectly balanced roulette wheel that contains alternating red and black slots and is set to spinning. The croupier (person who runs the roulette game for the management) throws a perfectly round ball into the wheel, against the direction of its rotation. Will the ball finally come to rest in a red slot or a black one? Which it does depends upon very small differences—too small for a human being to control—in the speed at which the wheel spins and the way in which the ball is thrown into the wheel.

Or think of the sperm traveling on its way to make the union with the ovum that created Napoleon. A small deflection from its path and that sperm would have been nothing. Another sperm would have accomplished the mission. There would have been no Napoleon, with whatever consequences that might have had for the history of the world.

Or finally think of George III, who was king of England at the time of the American Revolution. In America he was the object of bitter hatred, and his behavior was one of the things that made a continued tie to Britain intolerable. King George's tactless, hostile, unbending personality was a part of a set of symptoms so severe that he had been diagnosed as a "lunatic." Although King George's madness has sometimes been blamed on childhood experience, it now seems clear that it was actually the result of a physical condition, *porphyria*.

Porphyria is a genetic defect which makes it impossible for the body to metabolize a substance, porphyrin, in the hemoglobin of the blood. The substance accumulates and affects the brain, producing symptoms of mental disorder. It also gets into the body fluids, producing a classic symptom which King George had—wine-red urine.

The disorder is the result of an infinitesimally small genetic accident involving a single gene. Except for this tiny chance happening, the recent celebration in America might have been of the tricentennial anniversary of our membership in the British Commonwealth rather than the bicentennial anniversary of our status as an independent nation. As in the other examples, the point is that very large consequences are the result of very small causes. As Joan Beck put it in her newspaper column, where I first ran into the King George example, "For want of a nail...."[1]

[1]From a couple of comments on this example, I gather that the reference is no longer familiar to everyone. It goes as follows: "For want of a nail the shoe was lost. For want of the shoe the horse was lost. For want of the horse the rider was lost. For want of the rider the battle was lost. For want of the battle the war was lost. For want of the war the kingdom was lost. All for the want of a horsehoe nail." The point, as with the other examples, is that one feature of the happenings we call chance happenings is that infinitesimal causes have very large effects.

Multiple Causality

When someone tosses a penny into the air and allows it to fall to earth, whether it turns up heads or tails is certainly determined. In principle an intelligence or a computer that knew all the laws of mechanics and all the values to enter into the related equations could predict the outcome. But the amount of information that would have to be handled is unmanageably large. It would include force of the toss, pattern of rotation if the coin were spun, the influence of air pressure upon the coin, and the detailed features of the surface the coin lands on. The numbers involved would have to be very exact, and they would be very hard to obtain. Moreover, the effects of these influences would interact with each other. If two coins were flipped with different forces, this difference would multiply during the brief time before the coin came to rest. The differently spinning coins would be affected differently by the pressure of the air and they would almost certainly fall in different places with different surfaces. Even if they fell on exactly the same spot, they would behave differently then.

In the case of coin tosses, there is another point to consider. The behavior of the human operator who flips the coin is also determined by many variables. This adds to the complexity of the causal situation and it also tells us something important about these many causes. By and large they are uncorrelated or independent. For example, the different forces with which a person flips a coin each time probably do not depend upon the fine structure of the surface upon which the coin will fall.

Seeing this last point makes it possible, finally, to define a chance event, or in modern terms a "random event." *It is an event produced by many independent causes*. This definition obviously stresses Poincaré's second criterion of chance—multiple causality. The first criterion (large effects from little causes) calls attention to another point. Often in situations of chance the large effect is only large subjectively. Actually, there is a very small physical difference between the outcomes in roulette that find the ball in adjacent red or black slots. Psychologically, however, it is the difference between winning and losing—if the stakes are high, a very large difference indeed.

Principles of Probability

By now I hope that you will agree that I have thrown some light on the concept of chance. But *laws* of chance are something else. You would be quite justified at this point if you were to conclude that the argument seems to be moving us away from, rather than toward, an understanding of anything lawful.

If you will think back, however, you will recall that the discussion was always about individual random events. Such events are, in fact, so complexly determined as to be unpredictable and that makes them seem unlawful. The situation is very much different with large numbers of such events. And this is where the concept of *laws of chance* takes on meaning. The gambler cannot successfully predict heads or tails for the toss of a *single coin*, but for 1,000 tosses he can predict the number of heads with some accuracy. The mathematicians working for the insurance company cannot tell which *specific* individuals will die before the age of 65, but they can forecast how many people will. The opinion polster cannot know which *particular* citizens will vote democratic, but he can estimate the proportion with a small margin of error. To sum this idea up, and to anticipate what comes next, the laws of chance or *principles of probability* apply, not to single events, but to large numbers of them.

EVERYTHING YOU'VE ALWAYS WANTED TO KNOW ABOUT PROBABILITY

Suppose that someone tosses five coins into the air (or a single coin five times, the argument will be the same) and notes the number of heads and tails. For five *binomial events* (events with two possible outcomes), heads or tails for each, the outcome would have to be one of 32 possible sequences, sequences like THTHH and HHHTT. With a little patience you could work them out for yourself but they are all recorded at the bottom of Figure 5–2 (p. 92).

The first thing to understand about these 32 outcomes may not be obvious: everyone of them is equally likely. That is, TTTTT and THTHT and all the rest of these orders will occur in the long run with the same frequency, once in 32 tosses of five coins. The same thing is true of dice. In throws of two dice "snake eyes" (1-1) and "boxcars" (6-6), and the total of 7 produced *specifically* by a 3 on die number 1 and a 4 on die number 2 all have exactly the same probability, once in 36 throws for each outcome. Figure 5–1 shows all these possible results, together with some other information. Call the two dice individually die 1 and die 2. If die 1 shows a single spot, die 2 may show any number from 1 to 6. All these possibilities are shown in the extreme left-hand column of the figure. The second column from the left shows what can happen if die 1 turns up 2 spots. The next four columns show the remaining outcomes that can occur when die 1 turns up 3, 4, 5, or 6. The two outside columns give all the possible totals from 2 to 12 and the number of different ways in which each total can occur. The important thing to remember is that the total number of

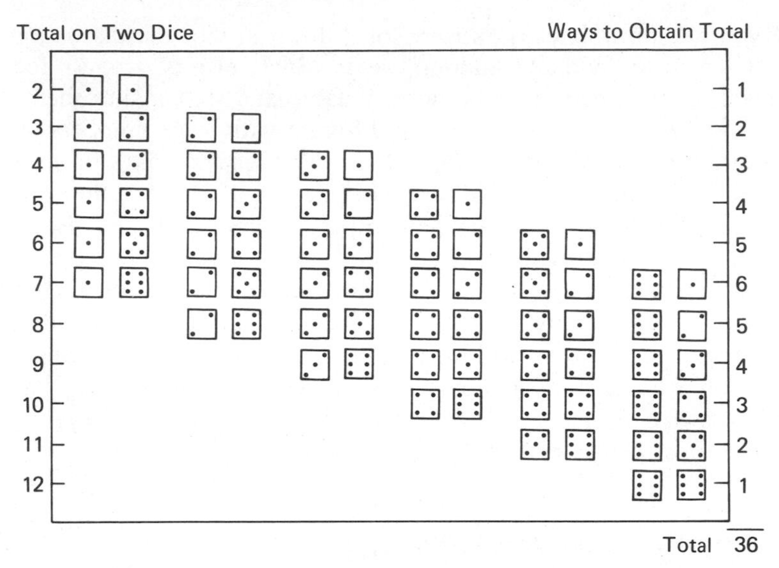

Figure 5–1 All 36 possible outcomes on a throw of two dice. The probability of any specific outcome is 1 in 36. The probability of throwing a combination whose total is 2–12 is the number of ways that outcome can occur divided by 36.

possible outcomes is 36 and that the probability of each separate outcome is 1/36.

One thing to notice about these two examples is that when I talk about the number of outcomes of a chance event I count different orders as different outcomes. Thus THTTH and HTTHT are different outcomes, although both orders consist of two heads and three tails. In the example of the dice, the orders 6-1, 5-2, 3-4, 4-3, 2-5, and 1-6 are all treated as separate outcomes, although each of them produces a sum of 7 spots showing on two dice.

In the probabilities involved in card games it is important to keep such considerations in mind. A 52-card deck will produce 311,875,200 five-card poker hands, if you count as different hands the 120 orders in which each five-card hand can be drawn. The number 120 comes about this way. For any particular hand the first card can be any of the five cards that eventually will make up that hand. The second, any of the remaining four; the third, any of the remaining three; and so on: $5 \times 4 \times 3 \times 2 \times 1 = 120$. In draw poker all of the 120 orders that produce a certain hand are equivalent[2] and the number you will see more often is that there are $311{,}875{,}200 \div 120 = 2{,}598{,}960$ possible hands in poker. Moreover, many of

[2]In stud poker this is not strictly true. I am tempted to illustrate this point here, but since I want to get on with the argument, I will refer you instead to my favorite book on poker: O. Jacoby, *On Poker* (New York: Doubleday, 1946). See page 270 in End Notes.

these 2.6 million hands are also nearly equivalent in terms of the likelihood of winning a pot. They are usually combined in the common tables of poker probabilities, like Table 5–1.

TABLE 5–1
Probabilities of Poker Hands

Hand	Ways to Make	P
Straight flush (5 cards in sequence, same suit)	40	.00002
Four of a kind (e.g., 4 aces)	624	.00024
Full house (three of a kind and a pair)	3,744	.0014
Flush (5 cards the same suit)	5,108	.0020
Straight (5 cards in sequence)	10,200	.0039
Three of a kind (e.g., 3 kings)	54,912	.0211
Two pairs (e.g., 2 aces and 2 queens)	123,552	.0475
One pair (e.g., 2 jacks)	1.098,240	.4226
"Nothing" (none of the above)	1,302,540	.5012
Total	2,598,960	.99996

As I shall explain more carefully in a moment, the probability of any random event is the number of ways the event can occur divided by the total number of possible outcomes. Calculations of that type produce the probabilities in the last column of Table 5–1. For poker hands these calculations are easiest to illustrate for "four of a kind." There are 13 such hands in a poker deck (4 aces, 4 kings, etc.) as far as just the 4 of a kind are concerned. But each of the 4 of a kind can have as a fifth card any of the remaining 48 cards in the deck after the 4 of the same denomination have been drawn. Thus there are $13 \times 48 = 624$ possible hands of 4 of a kind and the probability of getting one of them is $624 \div 2{,}598{,}960 = .00024$, or 24 times in 100,000 hands. Table 5–1 lists the hands in increasing order of probability, which you will notice is also the rank of the hands.

Distributions of Chance Events

Back to coin tosses. If I toss a single coin, there are two possible outcomes, H or T. If I toss two coins, there are four possible outcomes: HH, TH, HT, and TT. If I toss three coins, there are eight: HHH, HHT, HTH, THH, TTH, THT, HTT, and TTT. If you begin to detect an orderly series in the number of possible outcomes, this perception is correct. The numbers of possible outcomes are 2, 4, 8, 16, 32, etc.

Suppose, for the 8 outcomes produced by the toss of three coins, that I collect together all of those that have the same significance. If I do, I find that there is just one that produces three heads (HHH) and one that produces three tails (TTT). There are three outcomes that produce two heads and one tail (HHT, THH, and HTH) and three that produce two tails and one head (TTH, HTT, and THT). This accounts for all 8 outcomes. Figure 5–2 presents such information for tosses of coins ranging from 1 to 5 in number. I will use this presentation to introduce the ideas you need to have about probability.

The Probability of a Random Event. If you want to know about the probability of throwing five heads in a toss of five coins, you proceed this way. In throws of five coins, 32 different outcomes are possible. Only one of them produces five heads. Thus the probability is 1 in 32, or 1 ÷ 32

Number of Coins	Possible Combinations	Total Number of Outcomes
1	T (1); H (1)	1 + 1 = 2
2	T T (1); H T, T H (2); H H (1)	1 + 2 + 1 = 4
3	T T T (1); T H T, T T H, H T T (3); H T H, H H T, T H H (3); H H H (1)	1 + 3 + 3 + 1 = 8
4	T T T T (1); T T H T, T T T H, H T T T, T H T T (4); T H T H, H H T T, T T H H, H T H T, T H H T, H T T H (6); H H H T, H T H H, H H T H, T H H H (4); H H H H (1)	1 + 4 + 6 + 4 + 1 = 16
5	T T T T T (1); T T T T H, T T T H T, T T H T T, T H T T T, H T T T T (5); T H T T H, T H T H T, T H H T T, T T H T H, T T H H T, T T T H H, H T T T H, H T H T T, H T T H T, H H T T T (10); T H H H T, T H T H H, T H H T H, T T H H H, H T H H T, H T H T H, H T T H H, H H T H T, H H T T H, H H H T T (10); T H H H H, H T H H H, H H T H H, H H H T H, H H H H T (5); H H H H H (1)	1 + 5 + 10 + 10 + 5 + 1 = 32

Figure 5–2 All possible outcomes in throws of 1–5 coins.

=.03125. In more general terms, *the probability of any specified random event is the ratio of the number of ways that event can occur to the total number of possible outcomes.* Illustrating the concept with a different example, of the 32 possible outcomes that can occur in a toss of 5 coins, 10 will produce three heads. The probability of three heads, thus is $10 \div 32 = .3125$. Finally, you should note that the probability of throwing 0, 1, 2, 3, 4, or 5 heads is $32 \div 32 = 1.0$. The sum of the individual probabilities is certainty. *Some* outcome certainly will happen, $p = 1.0$.

The Either–Or Rule. Another way to look at the probability of obtaining three heads is this way. One way of obtaining this result is by the specific sequence, HHHTT. Its probability is 1 in 32—remember that specific outcomes are all equally likely. Another specific way, also with a probability of 1/32, is HTHHT. Figure 5–2 shows 8 additional ways to obtain three heads. Altogether this makes 10 ways of obtaining three heads and, as we just saw, the general definition of probability leads us to understand that the probability of three heads is 10 in 32. But the ideas just presented show you that this probability is also the sum of the probabilities of obtaining HHHTT *or* HTHHT *or* HHTHT on to the end of the list of 10 specific ways of obtaining three heads.

Again, in more general terms, *the probability of obtaining either one random event or another of a set of mutually exclusive random events is the sum of the individual probabilities*. The expression mutually exclusive in this definition of the *either–or rule* is very important. If you were to try applying the rule to poker hands asking, for three hands, about the probability that the first, second, or third hand would contain a pair, you would get (see Table 5–1) $.4226 + .4226 + .4226 = 1.2678$, an impossible probability, because the sum of the probabilities of all possible outcomes is 1.0. The reason that the *either–or rule* does not work is that pairs could occur in two or three hands, and this violates the provision of mutual exclusiveness. Ending the discussion with an example where the *either–or rule* applies, the probability of obtaining either 3 heads or 3 tails in a throw of 5 coins is $10/32 + 10/32 = 20/32 = .6250$. You will note that both three heads and three tails cannot occur. These outcomes are mutually exclusive.

The And Rule. Recall that the probability of throwing five heads is 1/32. A different way of getting to this conclusion introduces another important law of chance, the *and rule*. It goes this way. For the first coin the probability of a head is 1/2. For the second and all other coins, the probability is the same. The probability of five heads is $1/2 \times 1/2 \times 1/2 \times 1/2 \times 1/2 = 1/2 \times 2 \times 2 \times 2 \times 2 = 1/32$. In more general terms, for a last time *the probability of the joint occurrence of any number of random events is the product of their individual probabilities.*

Summary: Pascal's Triangle

The row of numbers at the bottom of Figure 5-2 (1,5,10,10,5,1) presents the number of outcomes that will produce zero heads, one head and four tails, two heads and three tails, and so on, for a toss of five coins. The row of numbers just above (1,4,6,4,1) presents corresponding information for tosses of 4 coins. The rows still higher in the figure are for throws of three, two, and one coins. The top part of Figure 5–3 is a summary of such materials for tosses of 1 to 10 coins. The figure is a portion of *Pascal's triangle*, named for Blaise Pascal, who invented it.

The numbers in the triangle itself are the numbers of different outcomes that contain so many heads and so many tails. For tosses of 10 coins (bottom row) you read the entries this way: one outcome produces zero heads, 10 produce 1 head and 9 tails, 45 produce 2 heads and 8 tails,...,252 produce 5 heads and 5 tails,..., 120 produce 3 heads and 7 tails,..., 1 produces 10 heads. If this is not clear, it will help to fill in the blanks above, to carry out the same exercise for other numbers of coins and to refer back to Figure 5–2, which contains the same information as is in the top five rows.

N		Total
1	1 1	2
2	1 2 1	4
3	1 3 3 1	8
4	1 4 6 4 1	16
5	1 5 10 10 5 1	32
6	1 6 15 20 15 6 1	64
7	1 7 21 35 35 21 7 1	128
8	1 8 28 56 70 56 28 8 1	256
9	1 9 36 84 126 126 84 36 9 1	512
10	1 10 45 120 210 252 210 120 45 10 1	1024
1	.500 .500	1.0
2	.250 .500 .250	1.0
3	.125 .375 .375 .125	1.0
4	.063 .250 .375 .250 .063	1.0
5	.031 .157 .312 .312 .157 .031	1.0
6	.016 .094 .234 .312 .234 .094 .016	1.0
7	.008 .055 .164 .273 .273 .164 .055 .008	1.0
8	.004 .031 .109 .219 .273 .219 .109 .031 .004	1.0
9	.002 .018 .070 .164 .246 .246 .164 .070 .018 .002	1.0
10	.001 .010 .044 .117 .205 .246 .205 .117 .044 .010 .001	1.0

Figure 5–3 Pascal's triangle in terms of possible outcomes and in terms of probabilities.

The column of numbers at the left is the number of coins—or any other binomial (two-valued) random event. The column at the right is the total number of possible outcomes—the sum of the numbers in the corresponding rows of the table. The entries within the two small inverted triangles show that successive rows in the triangle can be obtained by adding adjacent numbers in a given row and entering the total between them in the row below. Following this rule you could enlarge the triangle to any desired size by doing some minor calculations.

The bottom part of Figure 5–3 presents the probability of each of the entries in Pascal's triangle. Whereas the triangle gives the number of ways a specific outcome can happen, the version in terms of probabilities presents each number as a decimal fraction of the total number of possible outcomes. This presentation reviews the definition of probability 65 times —the number of entries in the bottom part of the figure—because the probability of an event is the fraction obtained by dividing the number of ways that event can be obtained by the total number of possible outcomes.

Figure 5–3 also makes it possible to review the meaning of the *either–or rule* and the *and rule* more times than you are apt to want to. Suppose that you would like to know the probability of obtaining five or more heads in a throw of 10 coins. This probability is that of obtaining *either* 5 heads *or* 6 heads *or* 7 heads *or* 8 heads *or* 9 heads *or* 10 heads. The top part of Figure 5–3 shows you that the number of ways of obtaining these outcomes are, in order, 252, 210, 120, 45, 10, and 1. The total of these numbers is 638 and $638 \div 1{,}024 = .623$. In terms of probabilities, take from the bottom row in the bottom half of the figure $.246 + .205 + .117 + .044 + .010 + .001 = .623$, the same value. The *either–or rule* is that the probability of obtaining one random event or another is the *sum* of the individual probabilities.

The *and rule* is that the probability of obtaining all of any number of random events is the *product* of their separate probabilities. One way to illustrate the operation of this rule is to ask about the probability of obtaining runs of any number of heads (or tails) that you choose. This is the same as asking about the probability of obtaining a head on the first toss of a coin *and* the second *and* the third *and* so on. The probability on the first toss is .50. This value multiplied by the same probability on the second toss is .25, which multiplied again by $.50 = .125$. Continued multiplication yields the series .500, .250, .125, .063, .031, .016, .008, .004, .002, and .001. This, you will note, is the same as the diagonal in the bottom half of Figure 5–3, which gives the probability of obtaining $1, 2, 3, \ldots, 10$ heads in a single throw of that many coins. As I mentioned in the first sentence of the section, the probabilities are the same whether the coins are tossed all at once or one at a time.

A CHANCY DEMONSTRATION

When I talk about probability to students, I often begin with a demonstration, partly because the results give the topic concrete objectivity and partly because they give me another chance to illustrate the logic of hypothesis testing. I begin the demonstration by making two predictions, which I write on a piece of paper and give to one of the students to read when the demonstration is over. The predictions are:

1. Your guesses will not be random.
2. Your guesses will show no evidence of extrasensory perception.

As this second prediction suggests, the demonstration takes the form of a little ESP experiment. I simply toss a coin 10 times, recording whether it comes down heads or tails on each trial, and ask the students to exercise their clairvoyance,[3] that is, to "see" which way the coin comes down. They record their guesses by writing down a series of "heads (H)" and "tails (T)." Finally, I ask each student to hand in two pieces of data: (1) the number of times they guessed "heads" in the first five trials, and (2) an ESP score, the number of "hits" or correct guesses. The reason for asking for number of heads on only five trials is that this allows me to make a test of these predictions with two different numbers of random events, five (guesses of heads) and ten (number correct).

Hypothesis Testing

If students are like coins and behave randomly in this little study, the distribution of their guesses is predictable from the probabilities in the bottom half of Figure 5–3. On the first five guesses the proportions of 0, 1, 2, 3, 4, and 5 heads should be .031, .157, .312, .312, .157, and .031. The proportions making $0, 1, \ldots, 10$ correct guesses should be $.001, .010, \ldots, .001$. These probabilities are those entered in the fifth and tenth rows of the table. The two null hypotheses to be tested are that the students will

[3]Investigators of paranormal phenomena study four main topics: *clairvoyance*, the perception of objects without the aid of the normal senses; *precognition*, the foretelling of things to come; *telepathy*, or mind reading; and *psychokinesis* ("mind over matter"), the mental control of physical objects. Along with most scientists I have an open mind as to the reality of these phenomena. If they exist, however, I believe that they must result from natural powers that have not yet been discovered. It is at least thinkable that we have sensory equipment beyond that studied in sensory physiology and that the electrical emanations from the body might influence physical events. What is not acceptable to me is the position taken by some people that extrasensory phenomena have no natural causes, that if telepathy is based upon some unknown physiological sense it would not be extrasensory and would, therefore, be of no interest. This position in effect denies the empirical basis of scientific knowledge and has no place in science.

distribute themselves in ways that are no different from these two expected distributions.

Guessing Data. The symmetrical distribution in Figure 5–4 shows probabilities just mentioned on the occurrence of 0, 1, 2, 3, 4, and 5 heads. The less regular distribution shown in dashed lines is the distribution of guesses for 100 students in the experiment described at the beginning of this section. Obviously, the distribution of actual guesses looks different from the theoretical distribution of random events. There are too many students who guessed 3 heads and too few who guessed 4 or 5. My calculations (p. 234) show that the odds are more than 100:1 against obtaining this distribution of guesses on the basis of random departures from the distribution of probabilities shown by the symmetrical function. Thus, in this case, I reject the null hypothesis, as I said I expected to, in making the first prediction with which this section began.

The distribution of guesses in Figure 5–4 occurs because of something important about most people's conception of randomness. It tends to contain too much of an emphasis on disorderliness and too little on the fact that some random sequences are regular. In the guesses these students made, too many of the sequences were irregular sequences (such as HTHHT and TTHTH) and too few were regular sequences (such as TTTTT and HHHHH). This accounts for a part of the departure from truly random performance shown in Figure 5–4. In addition, 66 of the 100

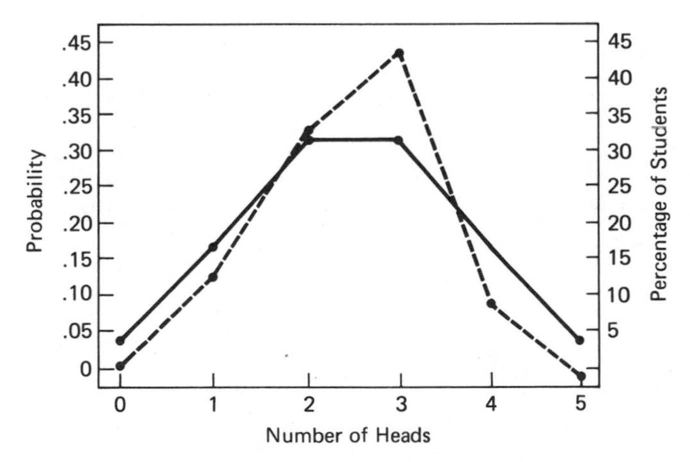

Figure 5–4 Comparison of the random distribution of numbers of heads with students' guesses. The abscissa is number of heads. The left-hand ordinate is the probability of each of these outcomes. The right-hand ordinate is the percentage of students who guessed each of these numbers of heads. As you can easily see, the students guesses are not random. There are too many guesses of 2 and 3 and too few of 0, 1, 4, and 5.

students guessed "heads" as their first guess. This, together with the bias toward irregularity, accounts for the fact that the distribution piles up at guesses of 3 heads.

In passing you should notice that my statement above that the odds are 100 : 1 against obtaining this distribution of guesses by chance means that we have another example of hypothesis testing. At the risk of boring you, let me review the steps, because they are important.

1. State the hypothesis in null form: the distribution of guesses is *no different* from expectations based on random probabilities.
2. Collect the required data. They are plotted in Figure 5–3.
3. Estimate the probability of obtaining the data if the null hypothesis is true. By a test called chi-square (p. 234) this probability turns out to be less than .01.
4. Accept or reject the null hypothesis. My calculations allow me to reject it at the 1% level of confidence.

Clairvoyance? The expected distribution of correct guesses in the ESP part of the experiment is the one shown in Figure 5–5. The points on the curve are the proportions taken from the bottom row of Figure 5–3. The unconnected points in Figure 5–5 are the percentages of students (right-

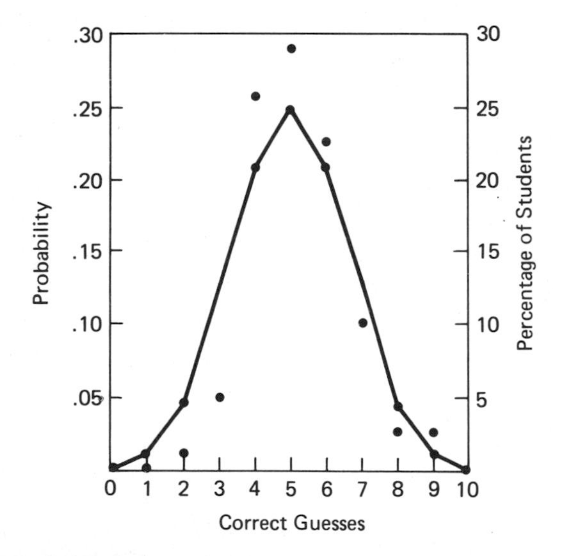

Figure 5–5 Test of clairvoyance. Solid lines show the expected distribution of correct guesses on the basis of chance. Solid, unconnected points show the distribution of guesses of 100 students. Statistical analysis shows that the students' distribution does not differ significantly from chance. Abscissa: number of correct guesses of heads or tails. Left-hand ordinate: chance probability of each number of correct guesses. Right-hand ordinate: percentage of students getting each number correct.

hand ordinate) getting each number correct. This time it is not obvious from the figure whether the data differ in any systematic way from expectation. A close inspection of the figure suggests, in any event, that the students did not show any very strong tendency to make large numbers of correct guesses. The most appropriate statistical test (p. 235) shows that these results would occur between 10 and 20 times in 100 replications of the experiment. This is too high a probability to provide a sufficient basis for assuming the existence of general clairvoyant power in this group of students.

But how about the five students who got ESP scores of 8 or 9? Have we, perhaps, located a small group who have unusual extrasensory talent? These questions lead to some of the most interesting points I have to make about the field of probability.

BINOMIAL PHILOSOPHY

The laws of chance begin by reducing to two alternatives, a particular outcome, P, with a probability, p, and all other possible outcomes, Q, which occur with a probability $q = 1 - p$. In the most important examples covered in this chapter p and q both have a probability of .5. That is, *on the assumption of randomness*, heads and tails in coin tosses and right and wrong in guesses of heads and tails each have a probability of 1/2. In the roll of a single die, by contrast p for any particular number is 1/6 and q (the probability of some other number) is 5/6. I just mention this last point in passing. The ideas I wish to develop can be made in terms of situations where the probabilities are 50–50.[4]

The Law of Large Numbers

It is time now to call attention to another feature of Figure 5–5. The actual data points did not fall exactly on the curve of expectation. Such discrepancies can serve to illustrate an important idea. Suppose that the guesses of the students in the ESP experiment *are* random, a conclusion our data seemed to suggest. The *law of large numbers* tells us that the actual data will come closer and closer to expectations as the number of observa-

[4]In this case, again, I want to avoid mathematical discussion. For those of you with the background, however, it may be useful to offer the following reminder. What I have been talking about in this part of this chapter is the expansion of the binomial formula $(p+q)^n$, where p and q have the meanings defined in the text and n is the number of events. The coefficients obtained in this binomial expansion are the numbers of ways to obtain a given outcome. The sum of the coefficients is the total number of possible outcomes, and each coefficient divided by the sum of the coefficients is the probability of that particular outcome. Since $p+q=1$, $(p+q)^n$ also always equals unity.

tions increases. Often the law is put in terms of expected proportions. In the case of coin tosses this means that the obtained fraction of heads would come closer and closer to 1/2 as the number of tosses increases. Or, to put it another way, the difference between the obtained fraction and the expected fraction of 1/2 will approach zero as the number of tosses approaches infinity.

The Gambler's Fallacy and Runs of Luck. The same thing holds for the individuals who made very high ESP scores in our experiment. If their guesses were just lucky guesses on these 10 trials—that is discounting the possibility of true clairvoyant power—the law of large numbers tells us that their proportion of correct guesses should converge on .5 with more and more trials. This does not mean, however, that the three people who correctly guessed 9 outcomes correctly are now more likely to be incorrect on the next trial. If correctness is the result of chance, the outcomes on successive trials are independent. To think otherwise is to commit the *gambler's fallacy*. Usually this takes the form of a belief that a long run of bad luck must mean that the chances of a better outcome are increased on the very next hand of cards, throw of the dice, or whatever. Not so. If the game is honest and the laws of chance apply, the outcomes of the successive deals or throws are independent.

The gambler's fallacy is just one of two ways in which subjective opinion errs because of a failure to recognize the independence of successive random events. The other way is a widespread belief in runs of luck. *Streak theory*, we might call it. A very important context in which both of these errors appears involves people's expectations about the sex of the next child, given a particular set of boys and girls already in a family. Gary McClelland kindly supplied me with the data that make this argument.

McClelland described families of various sizes to his subjects, giving ratios of boys to girls that ran all the way from three more girls than boys in the family to three more boys than girls. Then he asked for an estimate of the probability that the next child born to the family would be a girl or a boy. The data were the percentage of subjects who estimated correctly that the chances of the next child's being a boy or a girl were about equal, these data appear in Figure 5–6, where it is clear that people tend to make the correct estimate when the number of boys and girls already in the family is the same. The more unbalanced the family membership, the less likely a person is to make the correct estimate.

For purely logical reasons the mistakes these subjects made had to be either examples of the gambler's fallacy or streak theory. If they said that the probabilities were not the same, the only alternatives are that the imbalance in family membership will either be enhanced by the birth of another child of the gender that already predominates (streak theory) or that it will be at least partially corrected by the birth of a child of the other sex (gambler's fallacy). Figure 5–7 shows the data from Figure 5–6 broken

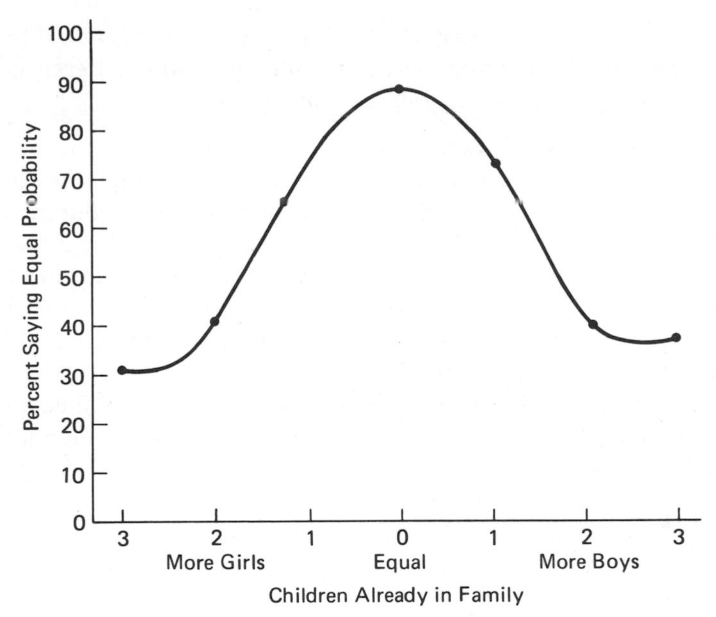

Figure 5–6 Failure to understand the independence of random events. The actual probabilities of boys and girls on a next birth are approximately 50–50, no matter what the makeup of the family. The well-known higher probability of a boy is *very* slight. People believe, however, that this is true only when the number of boys and girls already in the family is equal.

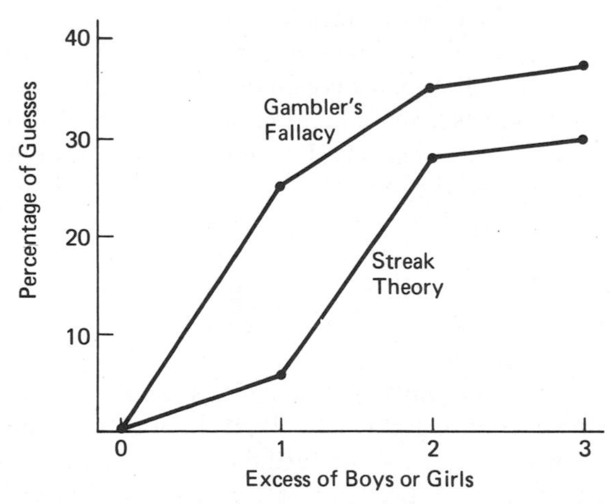

Figure 5–7 Two versions of the failure to understand the independence of random events. People divide themselves roughly equally into two groups. One group believes that, after a series of events of one kind, the "law of averages" dictates a change to the other event (gambler's fallacy). The other group believes in runs of luck (streak theory). The thing to note is the way these tendencies increase as the series of events of the same kind increases in number.

down in this way. No matter which of the errors an individual produces, the probability of the error increases with increasing differences in the number of older brothers and sisters in the family.

Lucky and Unlucky People. This way of thinking leads to ideas of more cosmic importance. Imagine that a distribution like the one in Figure 5–5 is for the numbers of good (lucky) and bad (unlucky) things that can happen to a person in life. If you think of heads as good things and tails as bad things, the translation into these terms is easy. Now consider what the form of the distribution means: it means that there always will be some people whom chance has dealt with kindly and others whom it has handled cruelly. They are the individuals in the two tails of the distribution. Our discussion of the gambler's fallacy tells us that there is no reason for such people to anticipate continued good luck or bad luck assuming that what is involved is truly luck. But judged by past happenings the lucky and the unlucky will always be with us. Or, as the philosopher C. S. Peirce does in one of his essays, consider the matter in time perspective. If we live long enough, all the bad things that can happen to us by chance will happen. Instead of that, we have death.

Extrasensory Immortality

Concepts of probability have sometimes been made a part of the basis for a belief in paranormal phenomena, the existence of a Supreme Being, or both. The reasoning can sometimes be persuasive. Suppose I dream that a friend will die in an automobile accident and then he does. Isn't this coincidence so unlikely that I *must* conclude either that a clairvoyant effect is at work or that God has taken a hand in the situation? The answer to the question is "no" for several reasons. In the first place as we just saw, very rare events do happen. In the second place, there is no real way to evaluate the probabilities that dreams will match actual happenings. In the third place, such interpretations ignore the negative cases, where dreams did not foretell anything. In the fourth place, when it is possible to compute the probabilities of coincidental events, those probabilities are often much higher than you might think.

Consider a favorite example of statisticians. Suppose that you have a party for 25 people, counting yourself. What do you think the chances are that two of you will have the same birthday? One in 25? One in 10,000? Those were the guesses of two people I asked just before I wrote the sentence. Actually, the chances are better than 50–50. Table 5–2 presents these probabilities for a selection of cases, where n is the number of people in the party. The calculation of these probabilities involves an application

TABLE 5–2
Probabilities That at Least 2 of n People Will Have the Same Birthday

n	p
2	.0027
3	.0082
4	.0165
5	.0275
10	.12
15	.25
20	.41
25	.57
30	.71
35	.81
40	.89
50	.97
60	.99

of the *either–or rule*. It gets a little intricate with large numbers of people, but there is nothing very mysterious about it, as I can show with a simple example. Take the case of 5 people and start with one of them. Ignoring the possibility of someone born on February 29 in a Leap Year, the chances are $1/365+1/365+1/365+1/365=.0027+.0027+.0027+.0027=.0110$ that one of the remaining four people will have the same birthday as the first person. Now move on to a second person, and figure the probability that one of the remaining three will have the same birthday as that individual. We consider only four people now because the preceding calculations have eliminated the first. They have also eliminated that person's birth date so that the probability of one of the remaining three people having the same birthday as the second person is $1/364+1/364+1/364=.0082$. Now move on to a third person and, using one less possible birthday, figure that the probability of a match is $1/363+1/363=.0055$. Finally, take the fourth and fifth people. The probability of a matching birth date is $1/362=.0028$. And now adding all these probabilities, $.0110+.0082+.0055+.0028=.0275$, the value in the table.

The Statistics of Tasting Beer. William H. Kruskal, a statistician at the University of Chicago, has described what turns out to be an interesting little puzzle. The example concerns an ad once used by the Miller Brewing Company, which went something like this:

> Many a "beer expert" has failed our *Miller High Life Taste Test*. We pour three glasses of Miller High Life, one from the bottle, one from the tap and one from the can and then we ask the "experts" which is which. So far not a

single person has identified all three correctly. The reason: all have the famous *Miller High Life* taste![5]

The point which the ad obviously wants to communicate is that Miller High Life packaged in these three ways cannot be discriminated. The fascinating fact of the matter is that the more data the company actually collected showing that no one makes the correct discrimination, the more certain it becomes that the point is wrong!

Look at the situation this way. There are only six ways in which the beers can be put into three glasses. The entire list follows:

Glass A	Glass B	Glass C
tap	can	bottle
tap	bottle	can
bottle	can	tap
bottle	tap	can
can	tap	bottle
can	bottle	tap

This list also identifies the six ways in which the taster can identify the three glasses. The point to understand is that, even if the taster cannot tell any difference at all among the beers, he should be right 1/6 of the time just by guessing and wrong 5/6 of the time.

At this point we can make use of the *and rule* to calculate the probability that taster 1 *and* taster 2 *and* taster 3 *and* ... and taster n will all be wrong. The probability is

$$5/6 \times 5/6 \times 5/6 \times \cdots \times 5/6$$

Here are the probabilities for several numbers of tasters.

1	$5/6$	=	.83
2	$5/6 \times 5/6$	=	.69
5	etc.	=	.39
10	etc.	=	.16
20	etc.	=	.02
30	etc.	=	.004

If you glance back at the ad you will notice that an important piece of information is missing—the number of people it takes to make "many a

[5]"Testing Beer Tasters," Chapter 11 in F. Mosteller, W. H. Kruskal, R. F. Link, R. S. Pieters, and G. R. Rising, *Statistics by Example: Exploring Data* (Reading, Mass.: Addison-Wesley, 1973).

beer expert." That is, we do not know the number of tasters who have failed the test. If the number is 5 or even 10, there is nothing too surprising about the data. Given the odds, the chances are 39/100 that all 5 tasters would be wrong and 16/100 that all of 10 would be wrong. If the number of tasters were as high as 20 or 30, however, it would be very unlikely (2/100 or 4/1,000) that the result would be obtained if the tasters were making no discriminations at all.

But what would this mean? It would mean that the tasters *were* making discriminations but identifying things wrong. If, for example, the tasters detected differences, but the can beer seemed more like what they thought tap beer should taste like, they would always call the real can beer tap beer and they could never be right. This would account for the reported results.

As a final thing, I hope that you will notice that this example tests the null hypothesis once more. We have asked: If beer experts have no capacity for distinguishing among beers, what is the probability that all n tasters would fail to make a correct identification? We have seen that with 20 tasters or more, the probability is very low. This, as usual, leads us to reject the null hypothesis, at the .02 level of confidence with $n=20$, and at the .004 level of confidence with $n=30$. The irony in the example is that the rejection of the null hypothesis results from the accumulation of data that individually support it.

One in a Billion

M. J. Moroney starts a discussion of the *and rule* this way.

> We shall now prove, to the no little satisfaction of the fair sex, that every woman is a woman in a billion. It is hoped that men-folk will find salve for their consciences in this scientific proof of the age-old compliment ("Statistics show, my dear, that you are one in a billion."). It will be obvious to the reader that the more exacting we are in our demands, the less likely we are to get them satisfied. Consider the case of a man who demands the simultaneous occurrence of many virtues of an unrelated nature in his young lady. Let us suppose that he insists on [Now I depart from Moroney to list the virtues that a European gentleman is said to insist on] sophistication in the living room, gourmet skill in the kitchen, the talents of a whore in the bedroom, and [returning to Moroney] a first class knowledge of statistics. What is the probability that the first lady he meets in the street will put ideas of marriage into his head? To know this probability we must know the probabilities for the several different demands. We shall suppose them to be known as follows:

probability of living room sophistication	.01
probability of gourmet cooking	.01
probability of sexual talent	.001
probability of a first class knowledge of statistics	.00001

> Multiplying the several probabilities together we find...that the probability of the first young lady he meets...coming up to his requirements is $p = .000,000,000,001$, or precisely 1 in an English billion.[6] The point is that every individual is unique when he is carefully compared, point by point, with his fellows.[7]

SUMMARY–GLOSSARY

The laws of chance apply to collections of random events. The concept of randomness crops up in many places in the study of statistics. We have already met this notion, and will again, in connection with sampling. A random sample is a sample in which every individual item and every combination of individual items has an equal chance of being selected. Another way to express the part about combinations of individuals is to say that the selection of one has no influence upon the probability of selecting any other. As far as selection goes, this is to say that the individual items in a random sample are independent of one another. The idea of independence is central to an understanding of the principles of probability.

Probability. The number of ways in which an outcome of interest can happen divided by the total number of possible outcomes.

Binomial event. An event with two possible outcomes. The toss of a single coin provides an example. The two possible outcomes are heads and tails.

Random event. One of a set of events that has the same probability of occurrence as any other member of the set. In a different way of looking at it, an event that is the result of numerous independent causes.

Either–or rule. The probability of occurrence of one of a set of mutually exclusive outcomes is the *sum* of the individual probabilities. Mutually exclusive outcomes are outcomes of which only one is possible: a coin

[6]For numbers above 1,000,000, the American and British systems are quite different. A partial comparison follows. Larger numbers are listed in many unabridged dictionaries under "number."

Number	American System	British System
1,000,000	Million	Million
1,000,000,000	Billion	Millard
1,(12 zeros)	Trillion	Billion
1,(15 zeros)	Quadrillion	Thousand billion
1,(18 zeros)	Quintillion	Trillion

[7]M. J. Moroney, *Facts from Figures* (Middlesex, England: Penguin Books, Ltd., 1951), p. 8.

cannot turn up both heads and tails; the sum of spots on the faces of two dice cannot be both 7 and 11. The sum of the individual probabilities of all possible outcomes is always 1.0.

And rule. The probability of obtaining a given number of chance outcomes is the *product* of the individual probabilities. If the probability that you are reading this sentence on Sunday is 1/7 and if the probability that any day is gloomy is 1/5, the probability that you are reading the sentence on a gloomy Sunday is $1/7 \times 1/5 = 1/35 = .03$ (approximately).

Pascal's triangle. A numerical diagram that shows, for binomial events, the number of outcomes, among all the possible outcomes, that will produce a result of particular interest. The diagram below is a fraction of the triangle. The numbers on the diagonal of the triangle give the number of events (*n*, see below) being considered. This triangle contains the numbers for only 1, 2, or 3 binomial events.

```
      1   1
    1   2   1
  1   3   3   1
```

The entries are the number of outcomes that produce specified combinations of, for example, heads and tails. Reviewing with just one example, four outcomes $(1+2+1)$ are possible in a toss of two coins. Two of them produce one head and one tail (HT or TH). Thus the probability of one head in a toss of 2 coins is $2 \div 4 = .50$.

p = probability of obtaining an event of interest

q = probability of not obtaining an event of interest: $q = 1 - p$

n = number of such events under consideration

Law of large numbers. As n (see above) increases toward infinity, the difference between the obtained value of p and the true value of p approaches zero. Later, a very similar idea comes up in connection with sampling: the larger the sample, the better a given statistic approximates the population parameter.

Gambler's fallacy. Failing to understand that random events are independent, the erroneous assumption that a long string of one kind of random events increases the probability of the alternative outcome. The common appeal to the "law of averages" is an example.

Streak theory. Again failing to understand that successive random events are independent, the mistaken belief that a long string of one kind of random event is likely to continue.

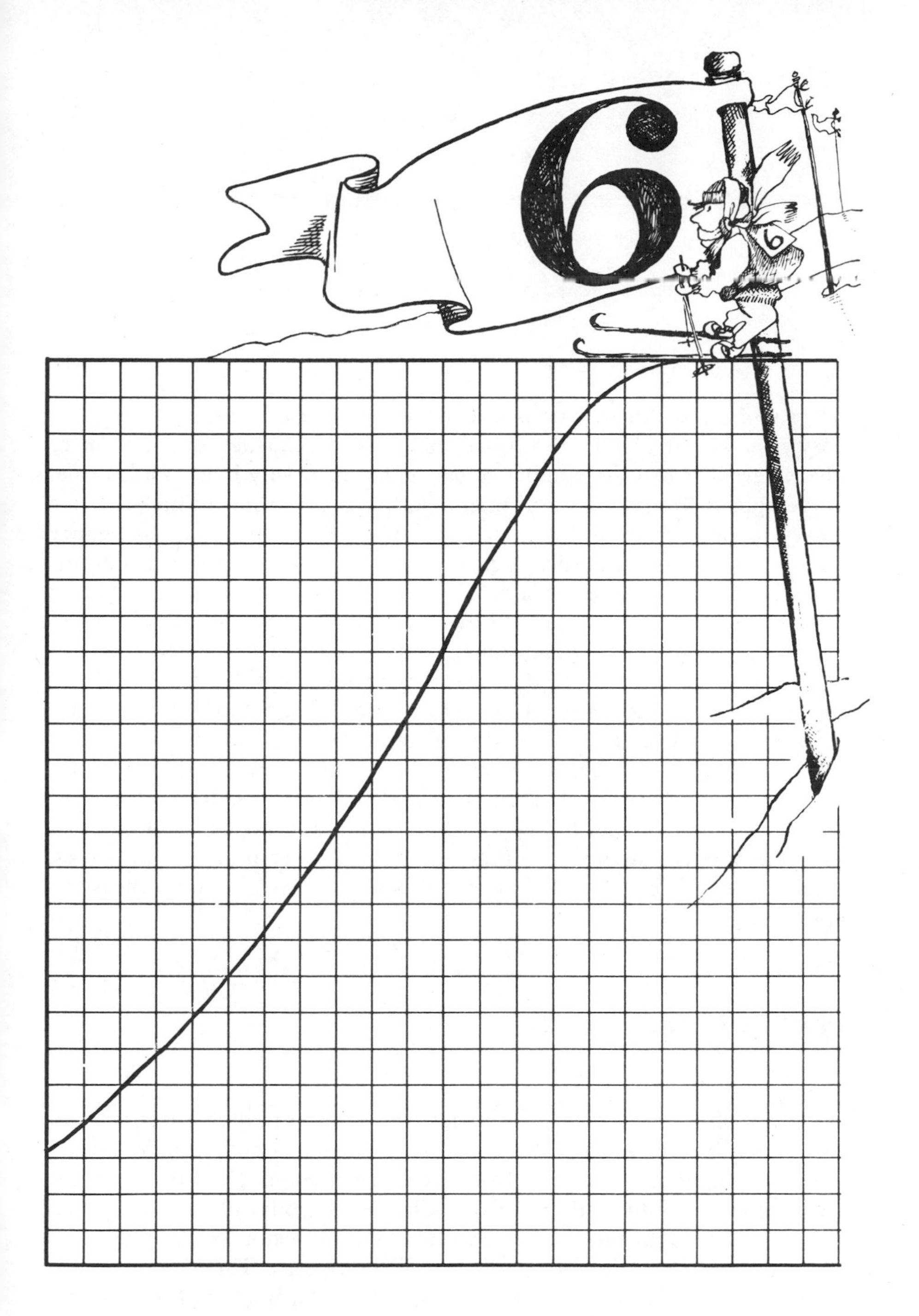

The Normal Curve

The step from the distributions of chance events to the normal curve is an easy one. Just think of the normal curve as a frequency distribution of chance events when the number of events becomes very large. The normal curve applies to many natural phenomena ranging from the sizes of tree leaves to the IQs of the human population. The connection to chance distributions means that such quantities are determined, as random events are, by many independent causes.

There is also another point to make about the examples mentioned in the last paragraph. The dimensions involved (sizes of leaves, measures of IQ) are continuous. That is, differences between things can be as small as the precision of measurement will allow. Matters are fundamentally different with random events. In tosses of coins, results are always in terms of whole numbers—so many heads and so many tails, never fractions. But a leaf can have an area of 12.7001 or 13.161942 square centimeters carried out to as many decimal places as you wish. To repeat the main point of this, the normal distribution function applies to continuous measures, whereas the distributions of chance events involve discontinuous measures and discrete events.

CENTRALITY AND SCATTER

As a way of leading up to a discussion of the important characteristics of the normal curve, I would like first to present a description of certain statistics that apply to any frequency distribution. If you recall the frequency distributions discussed in Chapter 3, you will remember that they showed a concentration of measures in one region of the graph but that the measures also spread out along the *X* axis. These observations identify the materials for discussion now, measures of *central tendency (averages)* and measures of *scatter* or *dispersion*.

Averages

The distribution of 101 annual incomes in Figure 6–1 approximates the actual figures for the late 1970s. I have presented the data with actual numbers in order to make it easy to describe three different types of average, the *mode* (most common score), the *median* (middle score), and the *mean* (what we all know as the average).

The Mode. The *mode* is the most frequent score in a distribution. As the data are plotted in Figure 6–1, the mode is in the salary range $9,000–12,000. If the data were plotted in finer units, however, the mode would be $26,000. There are three such salaries in the data, and no other occurs with a frequency greater than two. This shows at once why the mode is not a very useful measure of central tendency. It is too easily shifted by the accidental piling up of scores at some point which may be a considerable distance away from what obviously is the central tendency in a distribution.

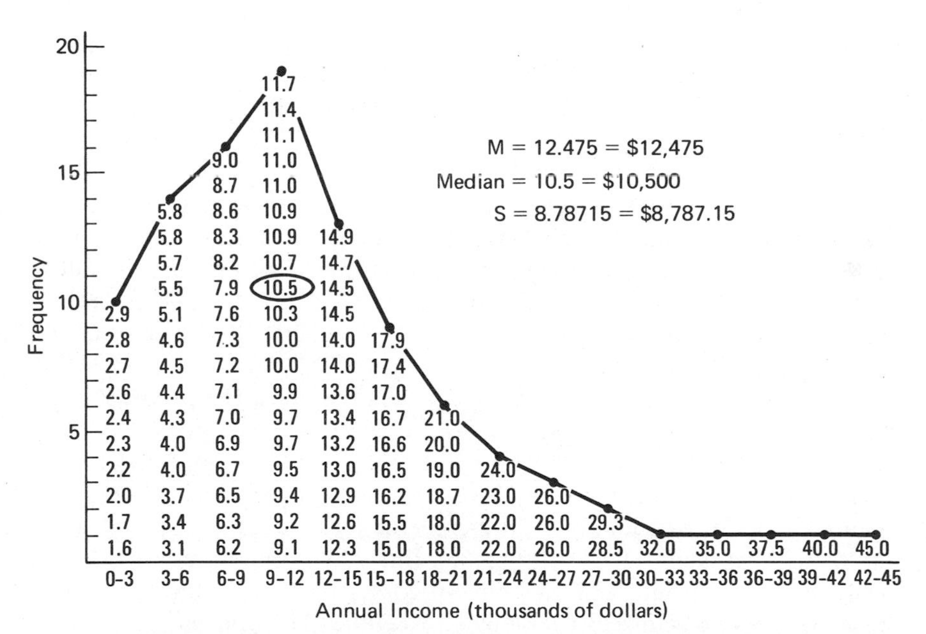

Figure 6–1 Approximate distribution of incomes. Several things to notice: The distribution is skewed. This makes the mean higher than the median. Although I have calculated the standard deviation for this distribution, you will learn later that *S* is not a very useful measure in the case of skewed distributions. At that point you might come back to this figure as a way of understanding this fact. Please note that salaries are divided by 1,000. This reduces *M* and *S* by the same fraction.

One place where the mode is a useful measure can be illustrated by a very famous set of data on the frequency with which cavalry men were killed by being kicked by a horse. There were data for 20 years and 10 corps of men, thus providing 200 observations. The data are in Table 6–1. Obviously the modal value of zero deaths describes the central tendency in this distribution quite well. The mode also describes quite well the typical conforming behavior of motorists at stop signs presented in Chapter 3 (p. 50).

TABLE 6–1
Deaths from Kicks by a Horse

Number of Deaths	Number of Years
0	109
1	65
2	22
3	3
4	1
5 (and more)	0
Total	200

The Median and Percentiles. The *median* is the middle score in a distribution. Returning to the data for annual salaries, since there are 101 salaries in Figure 6–1, the median would be the fifty-first salary counting either from the highest or the lowest salary. There are 50 salaries that are lower than the 51st salary and 50 that are higher. As it turns out, the median is $10.5 thousand.

Another way to put it is that the median is the fiftieth percentile, *a percentile being the percentage of scores which a given score equals or exceeds*. You would locate the salaries corresponding to other percentiles in a similar way. Take, for example, the 65th percentile. Since there are 101 scores in the distribution, the 65th percentile technically would correspond to the 65.65th salary ($101 \times .65 = 65.65$). But exactly such a salary does not exist in this distribution. This problem comes up more often than not in cases where one wants to compute a percentile. The convention is to say that the percentile in question is a fraction of the distance between two scores. In this case it would be .65 of the way between the 65th salary of $13,400 and the 66th salary of $13,600 or $13,530.

Turning the procedure for calculating percentiles around, you might want to determine the *percentile rank* of a salary of $4,400. Counting the number of salaries equaled or exceeded by $4,400, you find that it is the 17th score of 101 and thus has a percentile rank of 16.8 ($17 \div 101 = .168 \times 100 = 16.8$. A graphic way of dealing with percentiles is described in the computational appendix (p. 225).

The concept of percentile is one of a family of concepts that includes *deciles*, *quintiles*, and *quartiles*. Deciles divide a distribution into tenths; quintiles divide it into fifths; quartiles divide it into fourths. In Figure 6–1 the first decile (10th percentile) would be between the tenth and eleventh salaries ($2,900 and $3,100); the second decile would be between the twentieth and twenty-first salaries ($5,100 and $5,500). The second decile (20th percentile) is also the first quintile. Since quartiles divide a distribution into fourths, the first quartile (25th percentile) in Figure 6–1 would fall between salaries of $6,200 and $6,300. The third quartile (75th percentile) would be between salaries of $16,500 and $16,600.

The Mean. As you will remember, the computation of the *mean* requires the addition of all the measures to be averaged and then a division by however many measures there are. For the data in Figure 6–1 this would mean adding $1.6 + 1.7 + 2.0 + \cdots + 40.0 + 45.0 = 12{,}600 \div 101 = 12.475$, that is, $12,475. There must be a less cumbersome way to say all of this! There is, with the aid of a few symbols.

By convention, the letter X stands for an individual score. Similarly by convention, the capital Greek letter sigma (Σ) stands for the process of summation and our capital letter N stands for the number of measures. The availability of this symbolism makes it possible to describe the operations required to compute the mean (M) in this way:

$$M = \frac{\Sigma X}{N}$$

This little equation says what we already know: that to calculate the mean (M), the procedure is to add up (Σ) all the measures (X) and divide by the number of cases (N). For our example,

$$M = \frac{12{,}600}{101} = 12.475 = \$12{,}475$$

One thing that you will note about this value is that it is considerably higher than the median of $10,500 for this distribution. Such differences always occur with skewed distributions. The extreme scores in the longer tail of skewed distributions pull the mean in their direction. In the case of this positively skewed distribution, the mean is almost $2,000 higher than the median ($12{,}475 - 10{,}500 = 1{,}975$). If the distribution were negatively skewed (with the long tail toward the lower numbers), the mean would be lower than the median.

What the Mean Means. Suppose that there are 100 families in some community and that the distribution of numbers of children is as shown in the first two columns of Table 6–2. Six families have no children; nine families have 1 child; sixteen have 2; and so on to the one family with 12 children. Now suppose that you as a social planner decide that families with few children should help bear the burden of support with which families with many children are saddled.[1]

TABLE 6–2
The Sum of the Deviations from the Mean Is Equal to Zero

Number of Children, X	Number of Families, f	Number of Children × Number of Families, fX	d $(X-M)$	fd
0	6	0	−4.33	−25.98
1	9	9	−3.33	−29.97
2	16	32	−2.33	−37.28
3	14	42	−1.33	−18.62
4	13	52	− .33	− 4.29
5	11	55	+ .67	+ 7.37
6	9	54	+1.67	+15.03
7	7	49	+2.67	+18.69
8	5	40	+3.67	+18.35
9	4	36	+4.67	+18.68
10	3	30	+5.67	+17.01
11	2	22	+6.67	+13.34
12	1	12	+7.67	+ 7.67
Total	100	433		.00

Following up on this general notion, you decide that the fair way to proceed would be to charge small families $1 for each child fewer than the mean they have. In their turn the large families would receive $1 for every child more than the mean they have.

The mean[2] of the distribution of numbers of children is 4.33. Thus the six families with zero children would each be required to pay $4.33, because they have 4.33 fewer than the mean number of children. For the

[1]Given the population explosion, what I hope you actually believe is that it should be the other way 'round and that families with many children should be penalized while those with few children benefit. As I saw on a bumper sticker once, this might help to "Stop Heir Pollution." In discussions of statistical topics, however, the traditional meaning of the minus sign seems more important than social philosophy. And since minus signs customarily go with values below the mean, I set things up as I did. If you feel strongly on the social issue, you can reverse the signs and the results will be the same.

[2]Since the data are grouped in the table the calculation of the mean is easier if one uses the shortcut formula $M=\Sigma fX/N=433/100=4.33$. The computational appendix shows the calculation in this case in a little more detail (p. 223). The points I want to make here do not require you to follow the procedure, however.

six families the total is $25.98. Each of the nine families with 1 child (3.33 fewer than the mean) would have to pay $3.33, an aggregate of $29.97. On the receiving end of the distribution, the family with 12 children (7.67 more than the mean number) would receive $7.67 and the two families with 11 children would each receive $6.67, a total of $13.34.

Now let us see how our program for redistributing the wealth works out. The numbers required are to be found in the last column of the table. We collect $25.98 from the six families with no children. From this we can pay the $7.67 owed the family with 12 children and the $13.34 owed the two families with 11 children and have $4.97 left over.

$$\$25.98 - \$7.67 - \$13.34 = \$4.97$$

This will not pay the $17.01 owed the families with 10 children, however, and we must collect the $29.97 required of the nine families with 1 child, giving us

$$\$4.97 + \$29.97 = \$34.94$$

From this we pay the $17.01 and have $17.93 left, not enough to pay the $18.68 to the families with 9 children. So we collect $37.28 from the 16 families with two children and for the moment we are rich.

$$\$17.93 + \$37.28 = \$55.21$$

With this we can pay the families with 9 children their $18.68 and those with eight children their $18.35.

$$\$55.21 - \$18.68 - \$18.35 = \$18.18$$

But we need $18.69 to pay the families with 7 children, so we collect from the families with 3 children and have

$$\$18.18 + \$18.62 = \$36.80$$

Now we can pay off the families with seven children and those with six.

$$\$36.80 - \$18.69 - \$15.03 = \$3.08$$

and collecting our final $4.29 from families with four children, we have

$$\$3.08 + \$4.29 = \$7.37$$

just enough to pay the families with 5 children.

This tedious little exercise makes the following important point about the mean: *it is the point in the distribution from which the sum of the deviations is zero*. If you add up all the negative deviations and all the positive deviations in the table separately you will find that the two totals are −116.14 and +116.14, the total of those above the mean just canceling the total of those below.

The mean is the only point in the distribution for which this is true, as the following simplified example will show. Consider the numbers 1, 2, 3, 4, and 5. The total is 15.0 and the mean is 3.0. The upper portion of Table 6–3 presents the deviations and the sum of the deviations from every point in the distribution. The total is zero only when deviations are from the mean. While we are at it, it may be a good idea to use these same simple numbers to make a point that will have considerable significance in a later discussion. *The mean is the point in the distribution from which the sum of the squared deviations* (d^2) *is minimal.* The lower portion of Table 6–3 shows this for the numbers 1, 2, 3, 4, and 5.

TABLE 6–3
Sums of Deviations and Squared Deviations

	Deviation from X					
Number, X	X-1	X-2	X-3 (=X-M)	X-4	X-5	Total
1	0	−1	−2	−3	−4	−10
2	+1	0	−1	−2	−3	−5
3	+2	+1	0	−1	−2	0
4	+3	+2	+1	0	−1	+5
5	+4	+3	+2	+1	0	+10
Total	+10	+5	0	−5	−10	

Squared deviations from 1: 0+1+4+9+16=30
Squared deviations from 2: 1+0+1+4+9 =15
Squared deviations from 3: 4+1+0+1+4 =10
Squared deviations from 4: 9+4+1+0+1 =15
Squared deviations from 5: 16+9+4+1+0 =30

Measures of Variability

Table 6–4 gives the winners of the 100-meter run in 19 Olympic Games, together with the winning time in seconds. Just by inspection one can see that these times do not vary much among themselves. By calculation the mean winning time is 10.6 seconds and no time is more than 12.0 seconds or less than 9.9, the Olympic record.

The spread of times from 9.9 to 12.0 is the *range* of times. It provides one measure of the variability of this distribution but not a very useful one. It depends too much upon the single scores at the two ends of the distribution. The first winning time of 12.0 seconds makes this point. It is sort of off by itself, the next slowest time being 11.2 seconds. As with the

TABLE 6–4
Winners of the Olympic 100-Meter Run

Year	Winner	Time (sec)
1896	Thomas Burke (US)	12.0
1900	Francis Jarvis (US)	10.8
1904	Archie Hahn (US)	11.0
1906	Archie Hahn (US)	11.2
1908	Reginald Walker (South Africa)	10.8
1912	Ralph Craig (US)	10.8
1920	Charles Paddock (US)	10.8
1924	Harold Abrahams (Great Britain)	10.6
1928	Percy Williams (Canada)	10.8
1932	Eddie Tolan (US)	10.3
1936	Jessie Owens (US)	10.3
1948	Harrison Dillard (US)	10.3
1952	Lindy Remigino (US)	10.4
1956	Bobby Morrow (US)	10.5
1960	Armin Hary (Germany)	10.2
1964	Bob Hayes (US)	10.0
1968	Jim Hines (US)	9.9
1972	Valeri Borzov (USSR)	10.1
1976	Hasely Crawford (Trinidad)	10.6

mode we need to look for a more representative measure. Table 6–5 presents these data in a way that will suggest such possibilities.

In that table the times are listed in order from low to high and, in a second column headed d I have entered the difference between each time and the mean time of 10.6 seconds. Another way to refer to d, with the aid of the symbols presented earlier, is: $d = X - M$.

If we ignore the sign of d, the total of these differences or, better, deviations from the mean is 6.8[3] and the *average deviation* is .36 second ($6.8 \div 19 = .36$). This is one measure of variability that we might consider. I present it, however, only because the concept of d moves us toward an understanding of a much more useful measure of dispersion, the *standard deviation*.

The formula for the standard deviation (S) is this[4]: $S = \sqrt{\Sigma d^2 / N}$. I don't know that it helps much but, in words, this formula says that the standard deviation is the root-mean-squared deviation from the mean. You will need to use the formula now and then and I have worked out one

[3]If you do not ignore the sign, of course, the sum of the deviations is zero.

[4]*Note to the learned:* If you have a background in statistics, you may have seen a formula for S which has $N-1$ rather than N in the denominator. Those who use this formula do so because dividing by $N-1$ results in an estimate of the population standard deviation. My own opinion is that many things will be a lot clearer if I present these materials as I am doing it. In Chapter 7, when the question of estimating the standard deviation for the population arises, I will introduce the $N-1$ formula and explain it.

TABLE 6–5
Deviations and Squared Deviations from the Mean

Time	d	d^2
9.9	−.7	.49
10.0	−.6	.36
10.1	−.5	.25
10.2	−.4	.16
10.3	−.3	.09
10.3	−.3	.09
10.3	−.3	.09
10.4	−.2	.02
10.5	−.1	.01
10.6	.0	.00
10.6	.0	.00
10.8	+.2	.04
10.8	+.2	.04
10.8	+.2	.04
10.8	+.2	.04
10.8	+.2	.04
11.0	+.4	.16
11.2	+.6	.36
12.0	+1.4	1.96
Total	6.8 (disregarding sign)	4.26

example to show you how it goes. Note in the last column of the table of times for the 100-meter run that I have squared each deviation and have added them, obtaining a total (Σd^2) of 4.26. This value divided by $N = 19$ is .22. The square root of .22 is .47, the standard deviation of this distribution. In terms of our formula,

$$S = \sqrt{\frac{4.26}{19}} = \sqrt{.22} = .47 \text{ second}$$

What makes the standard deviation so important is its special relationship to the normal distribution function. I shall describe this relationship later in this chapter. For the moment I would like simply to stress the original point that the standard deviation is a measure of variability. The most important idea to have is that, whenever a distribution covers a small range of scores, S will be small. As distributions spread out more and more, S becomes larger and larger. One way to illustrate this is with values from some familiar distributions.

Weights and Bust Measurements of 17 Miss Americas. Seventeen of the women who recently have won the Miss America crown weighed in as follows: 114, 120, 116, 118, 115, 124, 124, 115, 116, 135, 125, 110, 121, 118, 120, 125, and 119 pounds. The mean of this distribution is 119.71 pounds and $S = 5.60$ pounds. Bust measurements for the same young women in the same order are: 34.5, 36, 35, 35, 36, 35, 36, 36, 36, 36.5, 36, 34, 36, 36, 36, 36, and 36 inches, obviously less variable than for weight: $M = 35.65$ inches, $S = .66$ inch. Reflecting the smaller variability, S is much smaller for bust measurements than weight.

Variations in IQ. As is well known, the mean IQ in the American population is 100. This is because a great deal of work has gone into construction of IQ tests to make the mean come out that way. The standard deviation depends a bit upon the test used and the age groups represented. The value is usually close to 15, however, a number that I shall use in a couple of examples later in the chapter. The range of IQs is from near zero to slightly over 200.

Ages of Presidents. The youngest president in the history of the United States, Theodore Roosevelt, took office at the age of 42; the oldest, William Henry Harrison, was 68. The mean age was 54.53 years; the standard deviation, 5.75 years.

Perhaps these examples will be enough to make the general point. The standard deviation is large where the range of scores is large and small where the range is small. You may want to notice that, where the ranges are similar, the values of S for two distributions are also similar. The range of weights for the 17 Miss Americas was 25 pounds ($S = 5.60$). The range of ages for presidents was 26 years ($S = 5.75$).

THE NORMAL DISTRIBUTION

As you will recall from the materials presented at the beginning of the chapter, the normal curve is a special frequency distribution that describes the distribution in the population of many continuously variable biological traits. For a sample of data, the formula for the best-fitting normal curve is

$$Y = \frac{1}{\sqrt{2\pi S^2}} e^{-\frac{1}{2}\left(\frac{X-M}{S}\right)^2}$$

For everyone except a sophisticated mathematician or a well-programmed computer, this equation is bad news because of its complexity. I present it only in order to make two important points. First, like any other equation, this one tells how Y changes with X. Specifically, this equation says that the Y is greatest at the mean, M, of X and that it falls in both directions from the mean approaching, but never reaching, the base line. Second, if you look at the equation carefully, you will see that exactly how Y changes in X depends upon only two terms that can vary, M and S. Everything else is either a constant (e.g., π and e) or it signifies a numerical operation (e.g., $\sqrt{\ }$). What this means is that if we know M and S for a given normal distribution, we know everything there is to know about that distribution.

To make all of this a little less abstract, suppose that we know that $M=4.0$ and $S=1.0$ for a certain distribution. Suppose further that with these terms available, we solve the equation for $X=1$, 2, 3, 4, 5, 6, and 7. The results[5] will look like this:

X	1	2	3	4	5	6	7
Y	.004	.054	.242	.399	.242	.054	.004

In order to give these numbers meaning in terms of the normal curve, I need to introduce you to one slightly new concept. Up to this point I have used a concept of *the* ordinate as the Y axis in a graph. Now I would like you to generalize this concept and think of *an* ordinate as the vertical distance from any point on the X axis to the associated value of Y. This means that the table above gives you the heights of seven ordinates. With the values of X as subscripts: $Y_1=.004, Y_2=.054,\ldots, Y_7=.004$.

If you have trouble with this idea, Figure 6–2 should clear things up. I have plotted a continuous normal distribution there but, at appropriate points on the base line I have erected these 7 ordinates and entered the calculated values. What the equation for the normal curve does is to give you these values for ordinates as close together as you choose. Remember that the normal distribution function applies to data where the values of X are continuous.

[5]*Where those numbers come from.* Although most of the mathematics required to obtain the values of Y are beyond the purposes of this book, the computation of the value of Y at the mean is so easy that I will do it, in this case, where $M=4.0$ and $S=1.0$.

$$Y=\frac{1}{\sqrt{2\pi\times 1^2}}e^{-\frac{1}{2}\left(\frac{4-4}{1}\right)^2}$$

$$=\frac{1}{\sqrt{2\times 3.1416\times 1}}e^{-\frac{1}{2}(0)^2}$$

$$=\frac{1}{\sqrt{6.2832}}e^{0}$$

And, since $e^0=1$,

$$Y=\frac{1}{2.5066}=.399$$

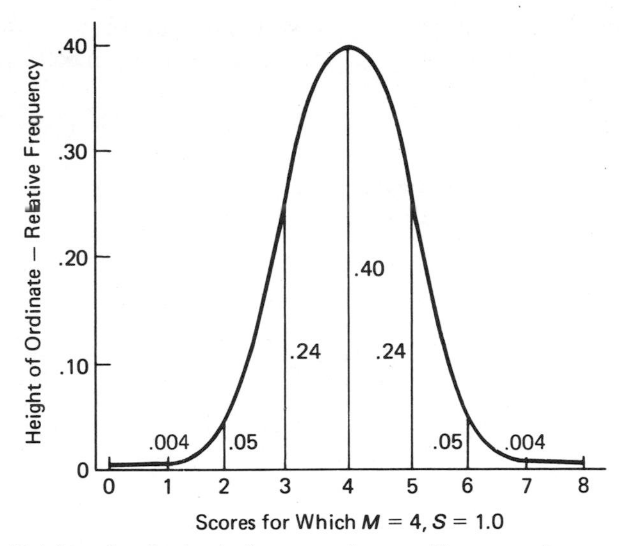

Figure 6–2 Heights of ordinates in the normal curve. These numbers are obtained by solving the equation for the normal curve where Y is the height of the ordinate at each value of X.

Areas Under the Curve

The most useful thing the equation for the normal curve does is to make it possible to calculate areas in different portions of the distribution. This, in turn, allows us to estimate probabilities. In order to show you how this works, you need to have the concept of Z scores.

Z Scores. The formula for the Z score is

$$Z = \frac{X - M}{S}$$

In words, *a Z score tells you how far above or below the mean any given score is in units of standard deviation*. It may be important to mention that you can compute a Z score for every score in a distribution.

To spell out the meaning of Z scores, let us consider the distribution of IQ scores, which are normally distributed[6] with a mean of 100 and a

[6]This is true if we ignore certain facts: (1) There is a group of mentally retarded people whose very low intelligence results from neurological damage, birth injury, for example. Including them makes the number of low IQs slightly too high. (2) Because of the facts considered in the section on the law of large numbers (p. 99), the distribution of actual data is not perfectly regular. (3) Since negative IQs are impossible, the distribution cannot be asymptotic to the base line at the low end of the distribution. None of these effects is big enough to spoil the example, however. As some of you will know, the standard deviation of the IQ distribution is often given as 16, the value for the Stanford–Binet test. I use 15 more or less routinely, however, because it works out so neatly when we consider IQs in 5- and 10-point steps, as in Table 6–6.

standard deviation of about 15. Now let us consider the set of data appearing in Table 6–6. I shall use these numbers to make a good many points about Z scores.

TABLE 6–6
Selected IQs, Their Z Scores, and Percentile Ranks

IQ	Z Score	Percent with Lower IQs
55	−3.00	.1
60	−2.67	.4
70	−2.00	2
80	−1.33	9
85	−1.00	16
90	− .67	25
100	− .00	50
110	+ .67	75
115	+1.00	84
120	+1.33	91
130	+2.00	98
140	+2.67	99.6
145	+3.00	99.9

1. It might not be a bad idea to check one or two of the calculations of Z scores to make sure that you understand that part. For one example which I present as a guide, consider the Z score corresponding to an IQ of 80:

$$Z = \frac{X - M}{S} = \frac{80 - 100}{15} = \frac{-20}{15} = -1.33$$

From this example and the table, you will see that scores below the mean correspond to negative Z scores. Those above the mean correspond to positive Z scores.

2. I suggest that you use the table to turn the concept of Z score into terms that are *verbal*. A Z score of $-.67$ is .67 standard deviation below the mean. An IQ of 130 is two standard deviations above the mean ($Z = +2.0$). An IQ of 80 (the example above) is 1.33 standard deviations below the mean.

3. It will be useful to note the relationship in the table between Z scores and percentiles which you know about. Figure 6–3 shows this relationship graphically. On the base line of the graph I have marked off the Z scores corresponding to the IQs in the table. The ordinate is the percentage of the population with lower IQs from the third column of the table.

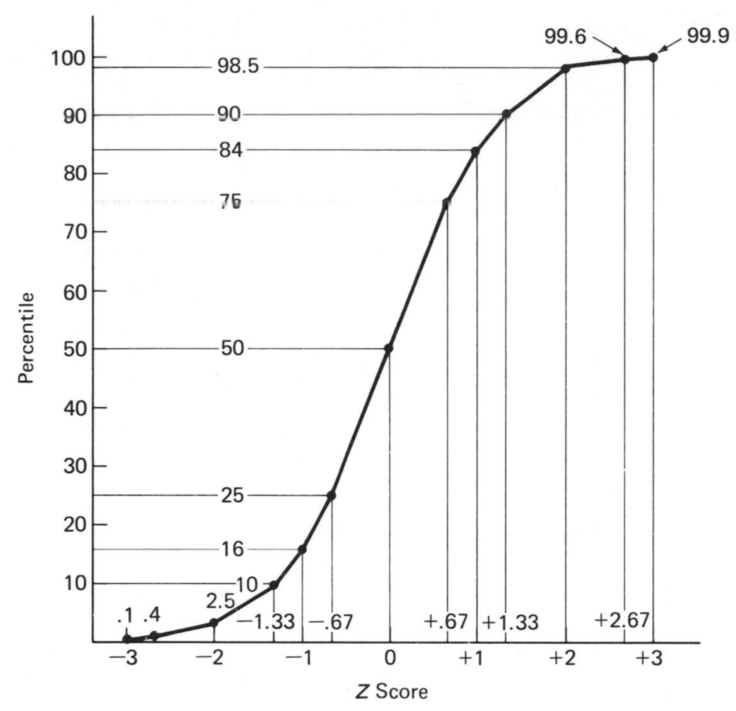

Figure 6–3 Cumulative percentage of area under the normal curve. In effect, this plots the percentile ranks in Table 6–6 against Z. If you were to start at the top of the last column in Table 6–6 and plot these numbers as a function of Z, you could reconstruct this graph.

Figure 6–4 makes some closely related points using the same numbers. This time, however, I have broken down the area in the two halves of the distribution in order to show the symmetrical proportions of area in the two tails of the distribution.

Figure 6–5 uses some of the same numbers in still a different way, this time to illustrate the percentage of cases (area) in a normal distribution bounded by symmetrically placed positive and negative Z scores. You should note that the range from minus to plus .67 embraces 50% of the area; the range from -1.0 to $+1.0$ contains 68% of the area; the range from -2.0 to $+2.0$ covers 95% of the area; and the range from -3.0 to $+3.0$ includes 99.7% of the area. These same values mean, of course, that less than 1% of the area in a normal distribution is beyond the range ± 3.0; 5% is beyond the range ± 2.0; and so on. These percentages will become very important in our discussion of the methods of hypothesis testing based on normal curve statistics.

Finally, Figure 6–5 makes a point that may already be obvious: *Z scores and all of the associated calculations related to area apply to any*

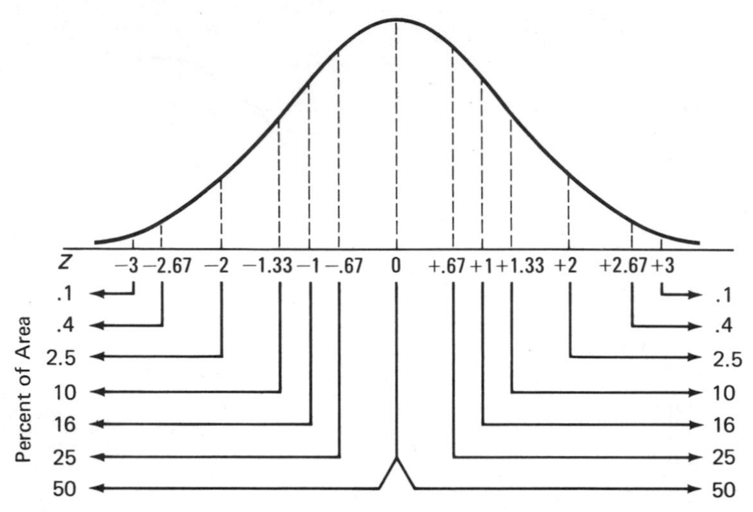

Figure 6–4 Percentages of area in each of the two tails of a normal distribution.

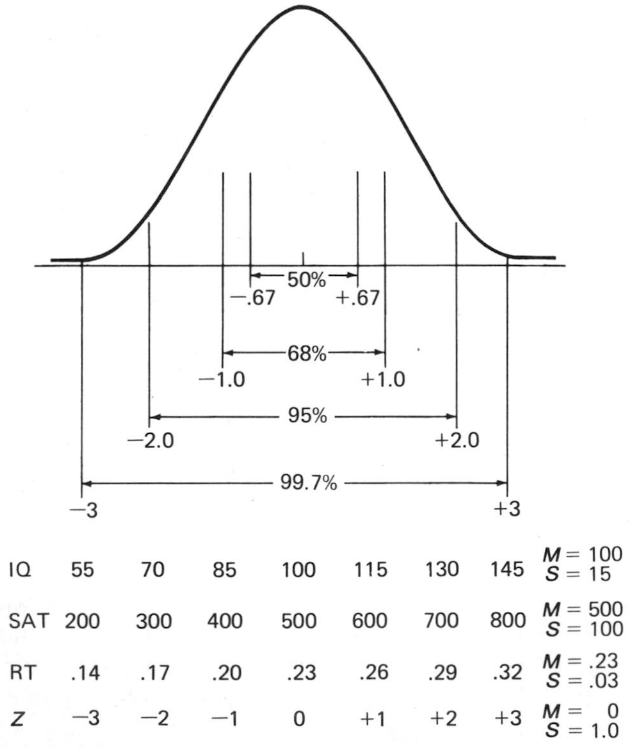

Figure 6–5 Areas bracketed by symetrically placed Z scores. The important point is that these values are the same for any normal curve.

normal distribution whatsoever, but only to normal distributions. The data I have used to make this point are IQs again, SAT scores which have (or did have until the national average began to decline) a mean of 500 and a standard deviation of 100, and some reaction-time data that I collected for the purpose of this presentation. *Reaction time* (RT) is simply the amount of time it takes to make a specified response to a particular stimulus—such as putting your foot on the car's brake when you see a red light. As you can imagine, the mean and standard deviation are both small, .23 second and .03 second, respectively, for these data. At the bottom I have entered points on the scale of Z scores for a final time.

TABLE 6–7
Areas in Normal Curve

Z Score	Area from Mean to Z	Area in Larger Portion	Area in Smaller Portion	Area from Plus to Minus Z	Area in Two Tails
.00	.0000	.5000	.5000	.0000	1.0000
.10	.0398	.5398	.4602	.0786	.9204
.20	.0793	.5793	.4207	.1586	.8414
.25	.0987	.5987	.4013	.1974	.8026
.30	.1179	.6179	.3821	.2356	.7644
.40	.1154	.6554	.3446	.3108	.6892
.50	.1915	.6915	.3085	.3830	.6710
.60	.2257	.7257	.2743	.4514	.5486
.70	.2580	.7580	.2420	.5160	.4840
.75	.2734	.7734	.2266	.5468	.4532
.80	.2881	.7881	.2119	.5762	.4238
.90	.3159	.8159	.1841	.6318	.3682
1.00	.3413	.8413	.1587	.6826	.3174
1.10	.3643	.8643	.1357	.7286	.2714
1.20	.3849	.8849	.1151	.7698	.2302
1.25	.3944	.8944	.1056	.7888	.2112
1.30	.4032	.9032	.0968	.8064	.1936
1.40	.4192	.9192	.0808	.8384	.1616
1.50	.4332	.9332	.0668	.8664	.1336
1.60	.4452	.9452	.0548	.8904	.1096
1.70	.4554	.9554	.0446	.9108	.0892
1.75	.4599	.9599	.0401	.9198	.0802
1.80	.4641	.9641	.0359	.9282	.0718
1.90	.4713	.9713	.0283	.9426	.0574
2.00	.4772	.9772	.0228	.9544	.0456
2.10	.4821	.9821	.0179	.9642	.0352
2.20	.4861	.9861	.0139	.9722	.0278
2.25	.4878	.9878	.0122	.9756	.0244
2.30	.4893	.9893	.0107	.9786	.0214
2.40	.4918	.9918	.0082	.9836	.0164
2.50	.4938	.9938	.0062	.9876	.0124
2.60	.4953	.9953	.0043	.9906	.0094
2.70	.4965	.9965	.0035	.9930	.0070
2.75	.4970	.9970	.0030	.9940	.0060
2.80	.4974	.9974	.0026	.9948	.0052
2.90	.4981	.9981	.0019	.9962	.0038
3.00	.4987	.9987	.0013	.9974	.0026

The main points to note about these numbers are the following:

1. The use of Z scores makes it possible to put any normally distributed set of data on the same scale. An IQ of 115 corresponds to a Z score of $+1.0$; so does an SAT score of 600 and a reaction time of .26 second. When we get to the topic of correlation we shall see that this makes it possible to ask about the relationships between scores—for example, is there any tendency for people with high (or low) IQs to have fast or slow reaction times? (The answer is "little or none.")
2. It is important to see that the Z distribution has a standard deviation of 1.0. All along we have been talking in terms of the total area of the Z distribution as 100%. If we talk in terms of decimal fractions instead, the area becomes 1.0. For this reason the distribution of Z scores is a *unit normal curve*. It has an area of 1.0, a mean of .0, and a standard deviation of 1.0. Table 6–7 shows how various Z scores divide this total area up into fractions in more detail than I have done so far. The table will be very important in our discussions of hypothesis testing.

I know from experience that, by now, many of you have begun to wonder how I know that the percentages of area recorded in Table 6–7 are what they are. This question is a mathematical question that is rarely answered even in fairly advanced statistics books, because it is difficult. The general form of the answer is easy to understand, however. If one *integrates* the equation for the normal curve between specified Z-score limits, the resulting values are those in Table 6–7. For those of you who are not enlightened by that statement, it will help to know that the mathematical process of integration is like addition. You can think of it as adding area between such limits.

SOMETHING OLD, SOMETHING NEW, SOMETHING BORROWED, SOMETHING BLUE

I begin with something borrowed.

As a sort of preview of what goes on in hypothesis testing with the aid of normal-curve statistics, consider the situation of one of the readers of the columnist Abigail Van Buren.[7]

Dear Abby:

You wrote in your column that a woman is pregnant for 266 days. Who said so? I carried my baby for ten months and five days, and there is no doubt about it because I know the exact date when my baby was conceived. My husband is in the Navy and it couldn't have possibly been conceived any other time because I saw him only once for an hour, and I didn't see him

[7]The following letter and response are reprinted by permission of Abigail Van Buren.

again until the day before the baby was born. I don't drink or run around, and there is no way that this baby isn't his, so please print a retraction about the 266-day carrying time because otherwise I am in a lot of trouble.

San Diego Reader

As Larsen and Stroup,[8] from whom I borrowed this example, comment: "Abby's answer was consoling and gracious but not very statistical":

Dear Reader:

The average gestation period is 266 days. Some babies come early. Others come late. Yours was late.

The Statistics of Pregnancy. What shall we make of this situation? One possibility, of course, is that the original letter was a hoax, concocted by some fertile undergraduate imagination at the University of California in San Diego. But investigation reveals that the situation as presented by San Diego Reader is not beyond the realm of conceivability. The article on pregnancy in *Britannica III* tells us that the typical gestation period is 266 to 270 days but that fully developed human babies have been born as early as, perhaps, 214 days after conception and that one court in New York, with medical advice, accepted a period of 365 days for purposes of establishing the legitimacy of a baby.

Suppose that we accept the authority of *Britannica* and assume that the mean gestation period is 268 days, the midpoint of the stated range of typical periods. In order to proceed with the statistical argument, we need to do three more things: (1) we need to accept the assumption that the distribution of gestation periods is a normal distribution; (2) we need to estimate the standard deviation of this distribution; and (3) we need to decide how many days there are in 10 months and 5 days. Given that we are dealing with a biological process, the assumption of normality seems at least reasonable. Various lines of evidence may be interpreted to mean that the standard deviation of the distribution is something like 14 to 16 days. Let us assume the highest value. As for the length of the gestation period reported by San Diego Reader, a fair estimate might be 308 days: 365 days in a year minus one month of 31 days minus 26 days in another 31-day month.

With all of these assumptions, the problem is this: In a distribution of measures where $M = 268$ and $S = 16$, what is the probability of occurrence of a measure of 308 or greater? The Z-score formula and a reference to Table 6–7 provide the estimate required.

$$Z = \frac{308 - 268}{16} = \frac{40}{16} = 2.50$$

[8]R. S. Larsen and D. F. Stroup, *Statistics in the Real World* (New York: Macmillan, 1976).

Consulting the table, we see that the area in the smaller tail of the distribution is approximately .0062. In short, about 6.2 births in 1,000 occur with a gestation period this long. To summarize the argument: *If* the mean human gestation period is 268 days, *and if* the distribution of gestation periods is normal, *and if* the standard deviation of this distribution is 16 days, *then* the probability of occurrence of a gestation period of 308 days is approximately .0062, or 6.2 times in 1,000.

Statistical Humanism. A probability value of .006 normally leads to the rejection of the null hypothesis. We conclude that such a probability is so small that it is implausible to believe that the obtained result is a chance happening. The particular case being considered, however, is very useful in pointing up some of the statistical and nonstatistical considerations that the making of such judgments involves.

1. In this case I am inclined to accept the account of her baby's conception offered by San Diego Reader, that is, to accept the null hypothesis. This is because to reject the null hypothesis has such dire consequences, at least for her—the stigma of adultery, the possible dissolution of a marriage. Such considerations always arise when one sets the *level of confidence* required to reject the null hypothesis. If the consequences are of great importance, it is natural to set a more stringent criterion than if the consequences are trivial. For scientific purposes, the 1 percent (.01) and 5 percent (.05) levels of confidence are the most widely used. In general, however, the more important the decision, the lower the probability value that will be set.

2. The logic of testing statistical hypothesis involves all of the assumptions (the "if" and "and ifs") that one must make to apply the test: *If* the mean gestation period is not 268 days *or if* the distribution of gestation periods is not normal *or if* the standard deviation of the distribution is not 16, *then* the calculated probability value of .006 could be very wrong.

In this case the assumption that is most suspect is the assumption of normality which had to be accepted to use normal-curve statistics. Recall, however, that the mean duration of pregnancies is 268 days. The shortest full-term pregnancy on record was 214 days, 54 days less than the average. The longest was 365 days, 97 days more than the average. The difference between 54 days and 97 days suggests that the distribution may be positively skewed (p. 48). If this departure from normality is extreme, it increases the probability of occurrence of long gestation periods. More important, it makes the application of normal curve statistics inappropriate.

The *general* point to understand is that the testing of statistical hypotheses always involves a set of supporting assumptions. If a null hypothesis is rejected because such a test reveals that outcome is highly improbable given those assumptions, it is important to recognize that the low probability could arise from a failure to meet the assumptions rather than from improbability of the result when the assumptions are true.

Something Old, Something New

This seems a good point at which to review some materials presented earlier in this book and also to enlarge your vocabulary a bit. We have just completed a small exercise in *hypothesis testing*. The hypothesis under test was the *null hypothesis*, specifically that a gestation period of 308 days is no different from what might be expected to occur occasionally by chance on the assumptions of a normal curve of gestation periods with a mean of 268 days and a standard deviation of 16 days.

The statistical test employed to evaluate the null hypothesis required the calculation of a Z score. For this reason the test is often called a *Z test.* The Z test is the granddaddy of tests of this type. When it appeared in the early statistical literature, it went by the name of the *critical ratio.* As we shall see later (p. 212), the Z test and critical ratio apply in an unbiased way to only when the standard deviation of the population involved is known. Otherwise, this value has to be estimated. This consideration is not very important with fairly large samples. When samples are small, a different, but closely related test, the *t test*, is more appropriate. With increasingly large samples, the values of t and Z become more and more nearly alike. Even when they are not alike, the logic underlying the use of these tests is always essentially the same. With this understanding you should be able to read a good bit of the literature in any area where reports of statistical tests play an important role.

This literature often contains expressions like this, which summarize the outcome of a statistical test: $t = 2.36$; $p < .05$. This phrase is to be read as follows: the t statistic was computed. Its value was 2.36. The probability (p) of obtaining a t of this size is less than .05. The null hypothesis can be rejected at the .05 level of confidence. Less frequently, you may run across such expressions as $t = 2.01$; $.05 < p < .10$. This notation tells you that the t value of 2.01 has a probability of occurrence that falls between .05 and .10 if the null hypothesis is true. Or, to put it another way, the level of confidence with which the null hypothesis can be rejected is between the 5% and 10% levels.[9]

[9]Often these reports, whether they are for t, F, χ^2, or any of a host of other statistics that you do not know about yet, will carry an indication of the number of *degrees of freedom* involved in the calculation. You may see $t_{30} = 2.01$ or $t = 2.01$, d.f. $= 30$, followed by a p value. I will introduce the concept of degrees of freedom later (p. 210). For the moment, it is enough to know that it is related to the number of observations. In the case of the t test, it is $N_1 + N_2 - 2$, where N_1 and N_2 are the numbers of individuals in two groups of an experiment.

As this last point begins to suggest, the concept of *level of confidence* is expressed in several ways. The values of p in the examples above are one way. Occasionally you may run into a use of the expression *alpha* level or just α, the lower case Greek letter A. It means exactly the same thing, as does the expression *significance level*, which you will also encounter in the literature. In a related expression you will sometimes encounter a reference to an *alpha error*, which means the same thing as what I have been calling a Type I error. All are terms for the probability of obtaining a statistic by chance if the underlying assumptions are true.

And, with Apologies, Something Blue

The only thing I can think of that will fulfill the promise implicit in the title of this section (Something Old, etc.) forces me to stoop to one of the obscure meanings of the word *blue* (characterized by indecency or obscenity). But maybe the following bit of doggerel will serve as an aid to memory.

The critical ratio is Z-ness,
But when samples are small, it is t-ness.
Alpha means A,
So does p in a way,
And it's hard to tell a-ness from p-ness.

SUMMARY–GLOSSARY

The normal distribution curve is a mathematical function whose form depends upon just two factors that can vary, the mean and the standard deviation. It describes the frequency distribution of many natural phenomena. In statistics this function is very often used in the testing of statistical hypotheses. In such applications, the logical argument always is: if the underlying distribution of phenomena of interest is normal with such and such values of the mean and standard deviation, the probability of obtaining the result being evaluated is so and so. The Dear Abby example in this chapter illustrated the procedure.

The important concepts in this chapter are the following:

Measure of central tendency. Any measure used to indicate the point in a frequency distribution where scores are concentrated. The mean, median, and mode are the examples discussed in this chapter.

Measure of dispersion. Any measure used to indicate the spread or scatter of a set of measures. Average deviation and range are two such measures barely mentioned in this chapter. The most important measures are the standard deviation and variance. (See below.)

Mode. The most frequent score in a distribution.

Median. The middle score or fiftieth percentile.

Mean (M). There are several ways to define it. (1) It is the arithmetic average. (2) It is the point in the distribution from which the sum of the deviations is zero, sign considered. (3) It is the point in a distribution from which the sum of the squared deviations is at a minimum. (4) The formula for the mean is

$$M = \frac{\Sigma X}{N}$$

where Σ indicates the process of summation (incidentally *not multiplication*), X stands for a score, and N = the number of scores.

Percentile rank. The percentage of scores that a given score equals or exceeds.

Standard deviation (S). The root-mean-squared deviation from the mean:

$$S = \sqrt{\frac{\Sigma d^2}{N}}$$

where for every score $d = X - M$, the difference between that score and the mean. The great value of the standard deviation is its relationship to areas under the normal curve.

Variance (S^2). Not covered in this chapter, it is the square of the standard deviation:

$$S^2 = \frac{\Sigma d^2}{N}$$

Later we shall see that variance also has an attractive property. It can be analyzed into additive components.

Z score. For any score, its distance above (+) or below (−) the mean in units of standard deviation:

$$Z = \frac{X - M}{S}$$

Z test. A test used to evaluate statistical hypotheses in which the Z-score formula is used to estimate the probability of occurrence of an obtained result if some specific hypothesis (usually the null hypothesis) is true.

Critical ratio. The Z test.

t test. A test like the Z test which is used when the standard deviation of the population is not known. Although it is not exactly correct, it is sometimes said that the t test rather than the Z test is used when evaluating hypotheses based on data from small samples. This way of putting it is not far wrong, because the differences between the t test and Z test are greatest when samples are small. This is covered more formally in the computational appendix.

p and probability. As used in this chapter, the probability of obtaining a result by chance if the null hypothesis is true.

Level of confidence. The same as p.

Alpha (α). The same as p. There is a tendency in statistical usage to set a value of α, an alpha level, that will be taken as an indicator of significance before analyzing results, and to use the expressions "level of confidence" and "p" or "p value" in discussions after the analysis is completed.

Alpha error. A Type I error. Rejecting the null hypothesis when it is true. As this should lead you to anticipate, the term "beta error" is sometimes used as an alternative to the expression Type II error, accepting the null hypothesis when it is false.

Sampling The Universe

Every year a certain mail-order house specializing in women's clothing had to face the question of which styles of dresses to advertise in their catalogs. In the past, decisions on this question had been left up to the advice of "experts." Unfortunately, this advice seemed always to fall somewhere on the dimension from inaccurate to disastrous. The experts simply could not forecast female taste in clothing. One year the company tried a different method. They prepared a preseason catalog containing several styles of dresses. This preliminary catalog went to a selection of potential customers and the final catalog carried the styles these women ordered most. Results were gratifying. These early sales did a better job of predicting which dresses women would buy than the experts had ever been able to do.

This example shows how data obtained from a *sample* can be used to estimate some important aspect of a *population or universe*. The customers who received the preliminary catalog were the sample; all the potential purchasers from the regular catalog were the population. In more general terms, a population is about what you would expect: all the members of a large group. A sample is a smaller selection from this group. The decisions involved in making statistical inferences are usually decisions about populations based upon information obtained from samples.

POPULATIONS AND SAMPLES

A population may be of finite size or it may be infinitely large. The population of a country counted in a periodic census can be very large, but it is not infinite. When we talked in Chapter 5 about the distribution of chance events in the very long run, we were talking about an infinitely large population. When the census taker reaches the last person in a country he cannot count any more people. The population is *finite*. Throws of dice and tosses of coins, by contrast, can go on forever. Such populations are *infinite*. Sampling theory as I shall present it in this chapter is usually about populations of infinite size.

Representativeness and Bias

Information from a sample can provide a valid basis for a decision about a population only if the makeup of the sample is like that of the population.

The Literary Digest Poll. The biggest fiasco in the history of sampling may have occurred in 1936. In that year the magazine *Literary Digest* mailed out 10 million straw ballots asking people to "vote" for Landon or Roosevelt, the Republican and Democratic nominees for President. More than 2 million of these ballots were returned. When the returns were published in *Literary Digest*, they predicted the election of Alfred M. Landon to the presidency. What happened in November, however, was that Franklin Delano Roosevelt won a landslide victory, receiving over 60% of the popular vote. The reason for the spectacularly inaccurate prediction of the *Literary Digest* poll was that the sample upon which the prediction was based was not a *representative sample* of the voting population. Ballots had gone out to subscribers to the magazine, to people listed in telephone books, and to owners of automobiles. Such a sample would include far too few poor people, who tend to vote Democratic.

Puzzle of the Pilfered Pantyhose. A company that made pantyhose once came to a firm of consulting psychologists asking for help in catching an employee who was systematically stealing a fraction of their product. The company had laid all sorts of traps but the criminal had not been apprehended. The evidence for thievery was that at the beginning of each month's production the company put its machinery into topnotch condition and did a sample run to see how many pantyhose the machines could produce with a given amount of raw material. Predictions based on these early data were always too high, hence the conclusion that some worker for the company was a crook.

Investigation produced quite a different explanation. When the company made its tests with the machinery freshly cleaned, oiled, and carefully adjusted, the equipment spun a finer thread and thus used less material to produce a given number of pantyhose. As the month wore on, the machinery lost some of this efficiency and produced fewer finished products than had been estimated from the sample. Again the trouble with the estimate was that the sample used to make a prediction (tests early in the month) was not representative of the population for which a prediction was being made (a full month's production).

Safety in Uncertainty

Samples that are not representative of the population of interest are said to be *biased*. The important thing to understand about biased samples is that the larger the sample, the more certainly the data will point to a wrong conclusion. The *Literary Digest* poll provides a beautiful example. Two million is an extremely large sample. But as the data from more and more unrepresentative people piled up, the more definitely they erroneously predicted a Republican victory. There is no safety in numbers as far as the consequences of selecting a biased sample are concerned. In fact, the situation is quite the opposite.

Randomness. The most straightforward way to avoid bias in a sample is to select it at random. *A random sample is one in which every individual and every combination of individuals has an equal chance of being selected.* A random sample is harder to obtain than this simple definition suggests. A few examples should serve to make the point.

1. There are some people who should have an equal chance of being selected in a random sample who wish to minimize this chance: criminals and people hiding from their creditors (or spouses), for instance.

2. Other people are hard to reach, for example, traveling salesmen, people on vacation, and working mothers. This decreases the probability of their being selected.

3. Sometimes it is impossible to tell whether a given individual is a member of the population of interest. In the case of polls of voters' preferences, for example, the population is the population of voters. But until a person casts his ballot (or fails to vote) it is impossible to say whether he is a member of this population or not. This is a major source of error in polls that try to predict the outcomes of elections. It may be worth mentioning that, in spite of this and other problems, the polls are more accurate than they have a reputation for being. Gallup's predictions for elections are usually accurate within a few percentage points. But a prediction of 49% Democratic/51% Republican, although quite accurate, is totally wrong in the sense that counts if the actual vote is 51% Democratic/49% Republican.

4. It probably does not happen with great frequency, but it is possible to make mistakes about the unit of sampling. Suppose that someone wants to determine the average number of children in the families in a certain school district and proceeds to ask all the children in the various classrooms how many brothers and sisters they have. Basing the estimate on responses to this question the pollster comes to the conclusion that the average number of children is 4.1, which is way too high. How could such a mistake have been made? There are at least two ways: (1) Large families would be more apt to be represented just because they are larger. If one of the children is absent from school because of illness, the large family will still be represented. (2) More importantly, the procedure as described allows large families conceivably to be counted as many times as they have children. Suppose that a couple entirely innocent of contraceptive information have 13 children evenly spaced from kindergarten to grade 12. This family would be counted 13 times when it should have been counted just once.

In this example it is important to note that a natural confusion is at the root of the problem. It is so common to think of *individual people* as the units to be represented in a sample that one might easily overlook the fact that in this case it is *families* that should stand an equal chance of being selected and mistakenly make a random selection of individuals.

STATISTICS AND PARAMETERS

Before I move on to the materials to be covered next, it will be important to make certain definitional distinctions related to samples and populations. Measures obtained on a sample are usually called *statistics*; comparable measures for a population are called *parameters*. Thus the mean and standard deviation of a sample are statistics; the same measures in the population are parameters. A common convention is to use uppercase Roman letters as symbols for statistics and lowercase Greek letters as symbols for parameters. Following this convention, I have been using M and S for the statistics obtained on samples. Now I shall use μ and σ for the population parameters. Although the interest of an investigation is usually in population values, the only information available is apt to be a set of statistics. An important part of inferential statistics involves the estimation of parameters from statistics. The cases we need to consider first involve the estimation of μ and σ.

The Estimation of μ from M

Provided that a sample is drawn at random, its mean is an unbiased estimate of μ. The only question that comes up is how good an estimate is it? An idea that you already have, the *law of large numbers*, tells you that this will depend upon the size of the sample. As sample size approaches the size of the population, M becomes a better and better estimate of μ, just as the obtained proportion of events approaches the true proportion (p. 99).

The Estimation of σ from S

As it turns out, S tends to underestimate σ for reasons that are important to present, at least briefly. In developing the first part of this explanation, it will be helpful to make the points in terms of S^2 and σ^2

rather than S and σ. The former terms, S^2 and σ^2, are symbols for the *variance* of a distribution, a concept that will take on a great deal of significance in the next chapter. Since the formula for the standard deviation of a sample is

$$S=\sqrt{\frac{\Sigma d^2}{N}}$$

the formula for the variance of a sample will be

$$S^2=\frac{\Sigma d^2}{N}$$

where d is the difference between each individual score and the mean, that is, $d=X-M$. Thus an alternative formula for sample variance would be

$$S^2=\frac{\Sigma(X-M)^2}{N}$$

The formula for population variance (σ^2), which is what we want to estimate, is slightly different:

$$\sigma^2=\frac{\Sigma(X-\mu)^2}{N}$$

where N is the number of individuals in the population and $X-\mu$ is the difference between a measure for each of these individuals and the population mean.

As we have just seen, M is an unbiased estimate of μ, but what must be stressed now is that M is almost certainly not exactly μ; it is in error by an unknown quantity $M-\mu$. On the average this means that $X-M$ in the formula for S is smaller than $X-\mu$ in the formula for σ by the difference between M and μ. In symbolic terms, $X-\mu=(X-M)+(M-\mu)$. This is why S^2 underestimates σ^2. Another way to put it is that, since M is the point in the distribution from which the sum of the squared deviations from the mean is minimal, Σd^2 will usually be less than the sum of the squared deviations from μ.

There is a straightforward proof[1] that the amount by which S^2 underestimates σ^2 can be expressed as follows. On the average,

$$S^2=\frac{N-1}{N}\sigma^2$$

[1]E. F. Lindquist, *Statistical Analysis in Educational Research* (Boston: Houghton Mifflin, 1940), pp. 48–50.

There are a couple of points to make about this formula: (1) you should note that, as N gets larger and larger, $(N-1)/N$ comes closer and closer to unity, which is to say that S^2 becomes a better and better estimate of σ^2; and (2) because of the precise way in which S^2 underestimates σ^2 [by a factor of $(N-1/N)$], it is possible to correct the estimate by multiplying S^2 by $N/(N-1)$. If we use the symbol $\hat{S}^2$ to stand for this corrected estimate,

$$\hat{S}^2 = \frac{N}{N-1} S^2$$

Substituting the formula for S^2 for the term itself,

$$\hat{S}^2 = \frac{\cancel{N}}{N-1} \frac{\Sigma d^2}{\cancel{N}}$$

$$= \frac{\Sigma d^2}{N-1}$$

and taking the square root of this expression, we get a corrected, or unbiased, estimate of $\sigma(\hat{S})$:

$$\hat{S} = \sqrt{\frac{\Sigma d^2}{N-1}}$$

SAMPLING DISTRIBUTIONS

Now to make an obvious but basic point, since M and $\hat{S}$ are estimates of μ and σ based on sample data, a different sample will usually give you a different estimate. I can illustrate some important consequences of this point by returning to our Dear Abby example for a moment. Suppose that for some reason, you wanted to make an independent estimate of the mean length of time between conception and the birth of babies, and proceed to do so. A cooperative obstetrician provides the data for his last 21 individual cases. Ordered from the shortest period of time to the longest, the number of days these 21 mothers carried their babies turns out to be as follows: 220, 230, 264, 265, 265, 265, 266, 266, 266, 267, 267, 267, 267, 268, 268, 268, 272, 278, 279, 280, 301. The mean and standard deviation of these data are 266.14 and 15.74, respectively. The value of $\hat{S}$ is 16.13.

But these data might leave you a little skeptical. You have always heard that the human gestation period is 9 months and that is more like 270 days than 266.14. Moreover, your faith in the regularity of biological processes and the precision of medical statements makes a population standard deviation of 16.13 seem very large. With your suspicions aroused, you decide to collect another sample of 21 gestation periods as a check. Suppose that the M and $\hat{S}$ for the second sample are 271.8 and 17.2 days. Still unconvinced you repeat the procedure—and continue repeating it, until you are either satisfied with the accuracy of your data or run out of obstetricians and patience.

You will by now have anticipated my main point: both M and $\hat{S}$ would be different from sample to sample. If you were to make a frequency distribution of all the means and standard deviations you had obtained, they would be sampling distributions for these two statistics. *The sampling distribution of a statistic is a theoretical frequency distribution of that statistic that you would obtain if you plotted the value of these statistics for a very large (actually infinite) number of samples.*

To put the concept of a sampling distribution in terms that are utterly concrete, suppose that you were able to collect 100 samples of 21 gestation periods. You obtain the mean of each and make a frequency distribution of these means. The result might be like what you see in Figure 7–1, which is an empirical sampling distribution of means. The distribution looks normal. Its mean (mean of the means) is 265.84 and its standard deviation is 3.82. The latter value is an estimate of what I shall introduce as the standard error of the mean (S_M) in the next section.

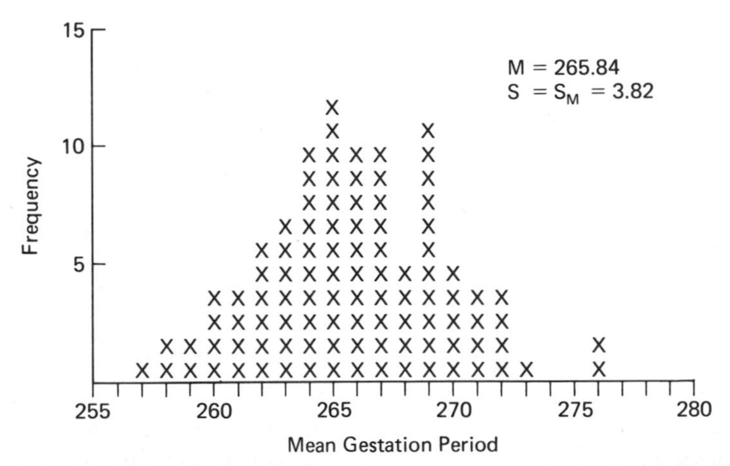

Figure 7–1 Empirical sampling distribution of the mean. Actually, these numbers are somewhat less imaginary than the text may suggest. In our example, the mean gestation period turned out to be 266.14 days. Later I shall estimate that the standard error of this mean is 3.54. To get the plotted values I took advantage of the powers of a programmable calculator. I set up a normally distributed population of means where $M=266.14$ and $S=3.54$. Then I drew 100 means at random from it and plotted the results. You will see that neither M nor S in the sample was exactly the population value of $M=266.14$ and $S=3.54$. This is how it is with sampling.

Operation Bootstrap

Obviously the example I have just used is unrealistic. In practice, it is usually impossible to collect sample after sample after sample in order to obtain a sampling distribution of the mean. More often the investigator has the mean of just one sample and must make do with it. He must pull himself up by his bootstraps, so to speak. While the discussion of the Dear Abby example is fresh in your minds, I will move the discussion forward a little by further discussion of it.

Brief Review. Recall that the mean gestation period of the first sample of 21 cases was 266.14 days. This is an unbiased estimate of the population mean (μ). The important point to keep in mind as the discussion proceeds is that 266.14 is just an estimate. Other samples would give other estimates, and many samples would provide a sampling distribution of the mean. What this review is leading up to is a way of estimating the standard deviation of the sampling distribution of the mean.

The standard deviation of the first sample of durations of pregnancies was 15.74 days, but this is not an unbiased estimate of σ. To obtain such an estimate, we must correct for the error. In order to summarize the previous presentation, it will be desirable to work with variance (S^2). The square of S, $(15.74)^2$, is 247.75. We multiply this value by the correction factor of $N/(N-1)$, $21/20 = 1.05$. Then $247.75 \times 1.05 = 260.14$, which is $\hat{S}^2$. The square root of 260.14 is 16.13, the unbiased estimate ($\hat{S}$) of the population standard deviation (σ).

The Standard Error of the Mean. The quantity we need to have for a variety of purposes is the standard deviation of the sampling distribution of the mean, S_M. For reasons that I will not explain, the formula for S_M is

$$S_M = \frac{\sigma}{\sqrt{N}}$$

We do not know the value of σ, but we do have an estimate of it, $\hat{S}$. Thus the working formula for S_M is

$$S_M = \frac{\hat{S}}{\sqrt{N}}$$

In the case of the 21 gestations periods, the calculations work out this way:

$$S_M = \frac{16.13}{\sqrt{21}} = \frac{16.13}{4.58} = 3.54$$

Remember that this is an estimated standard deviation of the sampling distribution of means. You interpret it exactly as you would any other standard deviation. The only difference is that the measures involved are means.The *standard error* of the mean is one of a class of closely related terms that mean very similar things. Whenever you see an expression that begins with the phrase, "standard error of ...," the translation is always the same. The "standard error of ..." is the standard deviation of the sampling distribution of whatever statistic completes the phrase. In this book we shall deal with the standard error of the mean, standard error of a difference, and standard error of an estimate. Respectively, these are the standard deviations of the sampling distributions of means, differences, and measures estimated from correlation coefficients.

Confidence Limits

Recall that the mean of the obtained sample of 21 gestation periods in our hypothetical example was 266.14. This, as we have seen, provides an estimate of the population mean. But how good is the estimate? We can be pretty sure that the true population mean is not exactly 266.14, if only because our hypothetical second sample gave a different estimate. Considerations suggested by this different estimate show that there is another way to look at the question. We can ask about the range of values within which the mean probably falls and state degrees of confidence that the mean falls within such ranges. Our estimate of the mean (266.14), together with the estimate of the standard error of the mean (3.54) and an assumption that the distribution of means is normal, are the tools that we need to describe the procedure.

Probably the most informative way to present the argument is, once more, with the aid of a graph, such as Figure 7–2. This figure shows the sampling distribution of the mean, estimated from the data on the single first sample. The mean of the distribution is 266.14 and the standard deviation (S_M) is 3.54. More important, the figure shows the ranges within which an obtained mean might be expected to fall 50% of the time, 95% of the time, and 99% of the time. These are the 50%, 95%, and 99% *confidence limits* for the mean.

All these ranges result from an application of knowledge you already have about normal-curve statistics. The confidence limits (sometimes called *fiducial limits*) are scores above and below the estimated mean that embrace the percentages in question. You may want to refer back to Table 6–7 (p. 125) to check your understanding. The 50% confidence limits are the mean plus or minus $.6745 \times 3.54$ (S_M). The 95% confidence limits are the mean $\pm 1.96 S_M$. The 99% confidence limits are the mean $\pm 2.58 S_M$.

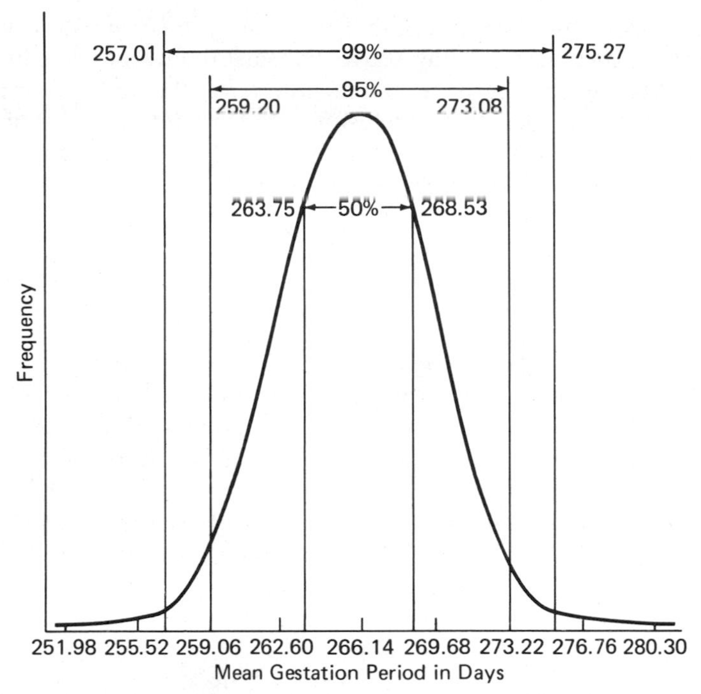

Figure 7–2 Theoretical sampling distribution of means. The figure is simply the normal curve where $M=266.14$ and S, which is $S_M=3.54$. The figure also shows the 50%, 95%, and 99% confidence limits for the mean. These were computed as follows: (1) the 50% confidence limits are $266.14 \pm .6745\ (3.54)$; (2) the 95% confidence limits are $266.14 \pm 1.96\ (3.54)$; and (3) the 99% confidence limits are $266.14 \pm 2.58\ (3.54)$. Two of these numbers—$M=266.14$ and $S=3.54$—should pose no problem. The other numbers—.6745, 1.96, and 2.58—are the Z scores that bracket the middle 50, 95, and 99% of the area in a normal distribution. That is, for example, the middle 50% of this area is in the range from $Z=-.6745$ to $Z=+.6745$.

Hypothesis Testing

Very similar ideas allow us to test specific hypotheses about the value of the mean. Since this is mostly review I shall not dwell heavily on the topic. But suppose one wanted to test the hypothesis that the mean human gestation period is 270 days and had the data we have been using to make such a test. The question becomes this: If the true mean is 270 days, what is the probability of obtaining a mean of 266.14 days? On the assumption that the distribution of means is normal, methods involving a Z score are appropriate:

$$Z = \frac{270-266.14}{3.54} = \frac{3.86}{3.54} = 1.09$$

A quick check of the last column in Table 6–7 (p. 125) will show that the probability of obtaining a Z far above or below the hypothesized mean is about .31. On this basis there is no reason to reject the null hypothesis. As it happens, we know that this conclusion is a Type II error: accepting the null hypothesis when it is false. Accumulated data on millions of pregnancies tell us that the true mean is less than 270 days.

ASSUMPTIONS

Things seem now to have progressed to the point where I can be more explicit about some matters that I have hinted at before. Inferential statistics, with the aid of a few numbers, is a process of making a certain kind of logical argument. As is apt to be the case with logical arguments, the statistical ones take the form: if such and such is true, then so and so follows. If you think back over the discussion of confidence intervals for the mean and hypotheses tested about the mean, you will see that these arguments rest on two *assumptions*.

The first of these assumptions is the assumption of randomness in the selection of the individuals in a sample. This assumption assures that M is an unbiased estimate of μ. In principle at least, the individual who does the sampling can determine whether this assumption is met. This is not true of the second assumption, which is that the sampling distribution of means is a normal distribution with a certain estimated standard deviation (standard error). Since statements about the confidence limits for a mean, or level of confidence with which an hypothesis is rejected, refer to the table of areas under the normal curve, these statements will be off to the extent that the actual distribution is not normal.

The Central Limit Theorem

The second of these assumptions brings us to one of the most important concepts in the area of statistical hypothesis testing, the central limit theorem. The *central limit theorem* is that the *totals (and therefore the means) of random samples will be normally distributed no matter what the distribution in the population is like, provided only that the samples are large enough.* Rather than attempting to prove this theorem abstractly, I will do it by example. Table 7–1 is a table of 900 random numbers. These days such a table is usually generated by a computer. However created, the essential feature of random numbers is that every digit from 0 to 9 stands an equal chance of appearing anywhere in the table.

TABLE 7–1
900 Random Numbers

	1	2	3	4	5	6	7	8	9	10	11	12	13	14	15	16	17	18	19	20	21	22	23	24	25	26	27	28	29	30
1	0	3	4	7	4	3	7	3	8	6	3	6	6	1	4	6	9	8	6	3	7	1	6	2	3	3	2	6	1	6
2	9	7	7	4	2	4	6	7	6	2	4	2	8	1	1	4	5	7	2	0	4	2	5	3	3	2	3	7	3	2
3	1	6	7	6	6	2	2	7	5	6	6	6	5	0	2	6	7	1	0	7	3	2	9	0	7	9	7	8	5	3
4	1	2	5	6	8	5	9	9	2	6	9	6	9	6	6	8	2	7	3	1	0	5	0	3	7	2	9	3	1	5
5	5	5	5	9	5	6	3	5	6	4	1	8	5	4	8	2	4	6	2	3	3	1	6	2	4	3	9	0	9	0
6	1	6	2	2	7	7	9	4	3	9	4	9	5	4	4	3	5	4	8	2	1	7	3	7	9	3	2	3	7	8
7	8	4	4	2	1	7	5	3	3	1	5	7	2	4	5	5	0	6	8	8	7	7	0	4	7	7	4	7	6	7
8	6	3	0	1	6	3	7	8	5	9	1	6	9	5	5	5	5	7	1	9	9	8	1	0	5	0	7	1	7	5
9	3	3	2	1	1	2	3	4	2	9	7	8	6	4	5	6	0	7	8	2	5	2	4	2	4	4	3	8	1	5
10	5	7	6	0	8	6	3	2	4	4	0	9	4	7	2	7	9	6	5	4	4	9	1	7	4	6	0	9	6	2
11	1	8	1	8	0	7	9	2	4	6	4	4	1	7	1	6	5	8	0	7	9	9	8	3	8	6	1	9	6	2
12	2	6	6	2	3	8	9	7	7	5	8	4	1	6	0	7	4	4	9	9	8	3	1	1	4	6	3	2	2	4
13	2	3	4	2	4	0	6	4	7	4	8	2	9	7	7	7	7	7	8	1	0	7	4	5	3	2	1	4	0	8
14	3	2	9	8	9	4	0	7	7	2	5	2	3	6	2	8	1	9	9	5	5	0	9	2	2	6	1	1	9	7
15	0	0	5	6	7	6	3	1	3	8	8	0	2	2	0	2	5	3	5	3	4	5	5	1	5	9	2	1	5	3
16	7	0	2	9	1	7	1	2	1	3	4	0	3	3	2	0	3	8	2	6	1	3	8	9	5	1	0	3	7	4
17	1	7	5	6	6	2	1	8	3	7	3	5	9	6	8	3	5	0	8	7	7	5	9	7	1	2	2	5	9	3
18	9	9	4	9	5	7	2	2	7	7	8	8	4	2	9	5	4	5	5	9	3	4	6	8	4	9	1	2	7	2
19	1	6	0	8	1	5	0	4	7	2	3	3	2	7	1	4	3	4	0	9	4	5	5	9	3	4	6	8	4	9
20	3	1	1	6	9	3	3	2	4	3	5	0	2	7	8	9	8	7	1	9	2	0	1	5	3	7	0	0	4	9
21	6	8	3	4	3	0	1	3	7	0	5	5	7	4	3	0	7	7	4	0	4	4	2	2	7	8	8	4	2	6
22	7	4	5	7	2	5	6	5	7	6	5	9	2	9	9	7	6	8	6	0	7	1	9	1	3	8	6	7	5	4
23	2	7	4	2	3	7	8	6	5	3	4	8	5	5	9	0	6	5	7	2	9	6	5	7	6	9	3	6	1	0
24	0	0	3	9	6	8	2	9	6	1	6	6	3	7	3	2	2	0	3	0	7	7	8	4	5	7	0	3	2	9
25	2	9	9	4	9	8	9	4	2	4	6	8	4	9	6	9	1	0	8	2	5	3	7	5	9	1	9	3	3	0
26	1	6	9	0	8	2	6	6	5	9	8	3	6	2	6	4	1	1	1	2	6	7	1	9	0	0	7	1	7	4
27	6	0	1	1	2	7	9	4	7	5	0	6	0	6	0	9	1	9	7	4	6	6	0	2	9	4	3	7	3	4
28	0	2	3	5	2	4	1	0	1	6	2	0	3	3	3	2	5	1	2	6	3	8	7	9	7	8	4	5	0	4
29	9	1	3	8	2	3	1	6	8	6	3	8	4	2	3	8	7	9	7	8	4	5	0	4	0	1	5	0	8	7
30	7	5	3	1	9	6	2	5	9	1	4	7	9	6	4	4	3	3	4	9	1	3	3	4	9	6	8	2	5	3

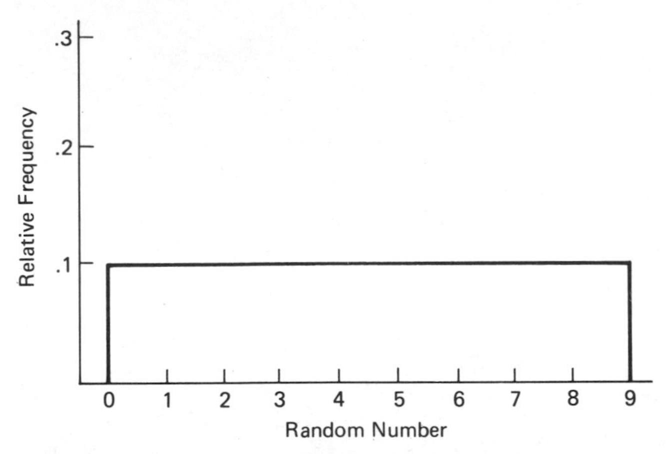

Figure 7–3 Rectangular distribution. Since the probability of any number's appearance in any position is .1 and the same for all numbers, a sample of a given size is expected to contain all the digits with equal frequency. The law of large numbers tells you that this is true only for infinitely large samples. The central limit theorem tells you that large enough samples drawn from this population will produce a normal distribution of means.

If you were to plot the number of occurrences of each digit from 0 to 9 in a frequency distribution, what would you expect that distribution to look like? Since the probability of each number is 1/10, you would expect it to be a rectangular distribution like that in Figure 7–3. You should realize, however, that the actual distribution would not exactly match your expectation. As with other distributions of random events, the actual distribution will only approximate the expected distribution until the number of events becomes infinitely large.

The mean of this expected distribution for the population is easy to determine: $9+8+\cdots+1+0=45/10=4.5$. For the standard deviation, you can either take my word for it that it is 2.87, or follow through the calculations in Table 7–2. These are appropriate because I have selected the one sample of 10 numbers that is exactly representative of the infinite population. As long as the number of 0s, 1s, etc., remained equal, S would still be 2.87.

The important thing about the situation in the table of random numbers is that we know the values of these parameters. The purpose of sampling is to estimate them. Since we do know these values in this case, we can go through the process of estimation and see how close we come to the known values. Table 7–3 contains the actual numbers drawn in 25 samples of 10 numbers each taken from a table of random numbers. The first sample contained the numbers 9005202750. The columns to the right of these numbers contain the calculated values of M, S, and $\hat{S}$.

TABLE 7–2
Calculating *S* for Numbers 0 through 9

	X	d	d^2		
	0	−4.5	20.25*		4.5
	1	−3.5	12.25		3.5
	2	−2.5	6.25	$S=\sqrt{\frac{\Sigma d^2}{N}}$	2.5
	3	−1.5	2.25		1.5
	4	−.5	.25		.5
	5	+.5	.25	$=\sqrt{\frac{82.50}{10}}$	.5
	6	+1.5	2.25		1.5
	7	+2.5	6.25		2.5
	8	+7.5	12.25	$=\sqrt{8.25}$	3.5
	9	+4.5	20.25		4.5
Total	45		82.50	=2.87	
Mean	4.5		8.25		

*Here is a little rule that may sometimes be useful: To square any decimal number where the decimal is .5, multiply the whole number by the next larger whole number and add .25. For example, $(3.5)^2=3\times4=12+.25=12.25$ and $(50.5)^2=50\times51+.25=2{,}550.25$. The same thing works for whole numbers; $(75)^2=7\times8=56$"+"$25=5{,}625$; $(45)^2=2{,}025$; etc.

You will note that these statistics vary from sample to sample.[2] Figure 7–4 is a frequency distribution of the values of these 25 means presented with a fitted normal distribution. This distribution of means is an actual empirical sampling distribution of means, like the one presented in Figure 7–1.

It will probably take a little or no argument to get you to agree that the distribution of means looks pretty normal. This fact provides an illustration of the central limit theorem. To repeat, the central limit theorem is that the distribution of totals (and therefore means) of random samples drawn from any population whatsoever will be normal, provided only that the samples are large enough. In this example the parent population was rectangularly distributed, but the distribution of means is already approximately normal with only 25 samples. We note incidentally that 10 numbers seem to satisfy the "large enough" clause in the central limit theorem.

[2]These data are actual data, drawn from tables of random numbers (here) or random normal numbers (to obtain the hypothetical distribution of mean duration of pregnancies). Although this way of approaching the topics of discussion is a little unusual, it appears to have the advantage of making the points concretely. In any event, please understand that none of this is faked. The procedures are as described and the statistics are what the HP 65 gave back to me.

TABLE 7–3

M, S, and $\hat{S}$ for Samples of Random Numbers

Sample	*Numbers*	*M*	*S*	$\hat{S}$			
					N	=	50
1	9005202750	3.00	3.13	3.30	M	=	4.36
2	8271827982	5.40	3.04	3.20	S	=	2.62
3	8934482877	6.00	2.37	2.49	$\hat{S}$	=	2.65
4	4444061944	4.00	2.32	2.45			
5	6610586843	4.70	2.57	2.71	N	=	100
6	1746779969	6.50	2.38	2.51	M	=	4.25
7	6767637401	4.70	2.45	2.58	S	=	2.83
8	5029046941	4.00	3.16	3.33	$\hat{S}$	=	2.84
9	7887019046	5.00	3.32	3.50			
10	9403874214	4.20	2.82	2.97	N	=	250
11	6012213542	2.60	1.80	1.90	M	=	4.46
12	2909802527	4.40	3.44	3.63	S	=	2.83
13	6705303497	4.40	2.84	2.99	$\hat{S}$	=	2.84
14	0076459007	3.80	3.34	3.52			
15	5431659097	4.90	2.88	3.03	N	=	300
16	0523871302	3.10	2.62	2.77	M	=	4.56
17	3644527181	4.10	2.30	2.42	S	=	2.86
18	9424935572	5.00	2.45	2.58	$\hat{S}$	=	2.86
19	6376375478	5.60	1.69	1.78			
20	2432774597	5.00	2.28	2.40	N	=	350
21	5265192644	4.40	2.24	2.37	M	=	4.49
22	6073282781	4.40	2.94	3.10	S	=	2.88
23	3325177045	3.70	2.24	2.36	$\hat{S}$	=	2.88
24	3324954613	4.00	2.14	2.26			
25	7624099180	4.60	3.47	3.66			
					N	=	400
					M	=	4.42
Mean		4.46	2.64	2.79	S	=	2.90
					$\hat{S}$	=	2.90

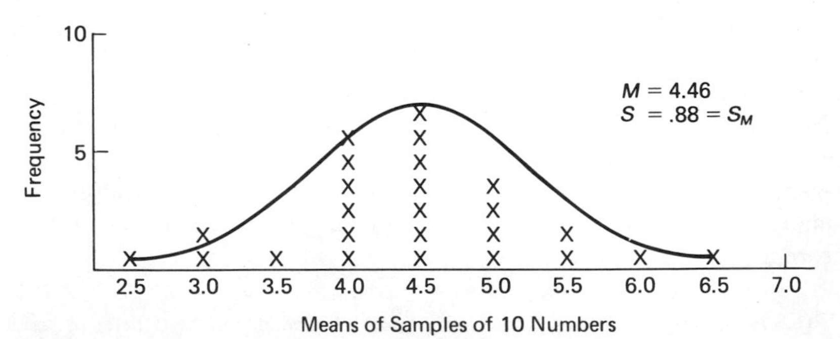

Figure 7–4 Sampling distribution of the means of 10 randomly selected random numbers. These data come from the table on this page. The mean as reported there is 4.46. The value of S of .88 was computed for the means. The smooth curve is the normal curve with these parameters. The important point is that, although the parent population was rectangular (see Figure 7–3), the distribution of means is normal. This illustrates the central limit theorem.

$\hat{S}$, the Estimate of σ

If you look at the means appearing at the bottom of the three columns in Table 7-3, you will notice two important things: (1) the mean of the means (4.46) is very close to the known value of 4.5, but (2) the mean value of S (2.64) is too small. If you examine the individual entries for S you will see that about two-thirds of them are smaller than the known value of 2.87.

There are three reasons for this. The first, as noted earlier, derives from the fact that the mean is the point in the distribution from which the sum of the squared deviations is minimal. As we saw then, there is a correction for this: using $N-1$ rather than N in the formula for the standard deviation where one wants to estimate σ. Such estimates for the 25 samples appear in the table in the column headed $\hat{S}$. A careful inspection of this column will show another thing, however. This revised formula still estimates an average value of σ that is a little too low (2.79 on the average rather than 2.87). This observation leads to a recognition of the second reason, mentioned earlier, for expecting too low an estimate of σ. The parent population is rectangular. Intuition will probably tell you that the absence of extreme values such as those occurring in the tails of a normal distribution will tend to keep the value of S, and therefore the estimate of σ, low. The third reason is practically the most important. The samples on which these estimates are based are small ($N=10$). This fact lies behind the development of the t test, which is used in place of Z for small samples. I discuss this test in the computational appendix (p. 235).

The materials on the right-hand side of the table show what happened when I drew larger single samples from a table of random numbers and computed the statistics of interest. As you can see, S and $\hat{S}$ remain too small until the sample size gets to be about 100. At that point the value of $\hat{S}$ is within .03 of the true value. After that the estimate remains close to 2.87.

SIGNIFICANCE OF DIFFERENCES

The step from a discussion of the sampling distribution of means to a similar discussion of differences is so small that you can take it almost without realizing that you have done so. The step is an important one, however, because it puts you in a position to understand the analysis of the results of experimental studies of the simplest type discussed in Chapter 3.

Somewhere in his *Principles of Psychology*, William James mentions, in discussing taste sensitivity, that he has known people who can distinguish between the top and bottom halves of a bottle of old claret (Bordeaux wine). For reasons that I will come to in a moment, it is virtually certain that these tasters would also have claimed that the bottom half of the bottle was superior. Because of my very high regard for this classical book, I remember that observation as one of James' lapses. It is almost certain that the ability he talks about exists, but that it has nothing to do with the gustatory powers of particular people. Two alternative hypotheses need to be considered. One is that the bottom half of the bottle finds people inebriated, and that they say that it is different and better in a spirit of good fellowship. I reject this hypothesis, for reasons related to the alcoholic content of Bordeaux wine, which is about 12%. If the top half of the bottle contains 12.5 ounces of wine, the amount of alcohol in it is exactly 1.5 ounces, the equivalent of two shots of 100 proof whiskey.[3] Now if James' people with the educated palates actually drank the whole top half of the bottle by themselves, this first explanation might be possible. But such behavior would have been unthinkable in the segment of polite society to which James belonged. Three or four people would have shared the bottle. We had better consider the second alternative.

This second alternative depends upon the fact that many wines, but especially those of Bordeaux, profit from "breathing," that is, exposure to air. For this reason, if you serve such a wine, it is wise to uncork it half an hour to as much as three hours before it is to be drunk. Since half an hour is not far from what it might take people to consume half a bottle of wine over a leisurely meal, the hypothesis that what is being detected is a difference due to breathing, and that almost anyone with a little wine-drinking experience would notice it, becomes a reasonable one.

Suppose that one wanted to test this hypothesis by experiment. There are several ways to do it. I shall present four of them and allude to a fifth in order to make some fundamentally important points. In doing so, I shall use the same numbers throughout. This will help to make the interpretive points come through with great clarity.

Tasters as a Random Variable, Between-Tasters Version

The most obvious way to do the experiment might be to select 20 tasters at random from the population of people who have enough acquaintance with wine tasting to rate wines on a widely used 20-point scale, where anything below about 10 points is essentially undrinkable and 20 points is perfection. For the experiment itself, these individuals would be divided into two random groups; one would taste the freshly opened wine, the other would taste it after breathing.

[3]The "proof" of a liquor is twice the percentage of alcohol. Many liqueurs are 86 proof, which means 43% alcohol. By law, in Britain the strongest drink available is 80 proof, or 40% alcohol. In this case 100 proof whiskey would be 50% alcohol. A shot glass contains 1.5 ounces, hence the conclusion that the top half of the bottle is worth two shots.

The next step would be to select a wine of decent quality, taking care, however, to select one that is not so excellent that it would always receive ratings of 19 or 20 even in its freshly opened state. Technically this is called a "ceiling effect": The scores under any condition are so high that differences cannot be detected. For wine too good, improvement with breathing might be so slight as to be undetectable. From there on, the procedure is straightforward. The two groups of subjects would taste the wine and give it a rating. Suppose the data for the two groups are shown in Table 7–4.

TABLE 7–4
Hypothetical Results of Wine-Tasting Experiment

Group A: Tastes before Breathing		Group B: Tastes after Breathing	
Subject	*Rating*	*Subject*	*Rating*
1	16.0	11*	16.0
2	9.5	12	12.0
3	15.0	13	15.5
4	15.5	14	15.0
5	13.0	15	14.0
6	16.5	16	18.5
7	14.5	17	18.0
8	17.0	18	13.0
9	12.0	19	17.5
10	17.5	20	17.0
Mean	14.65		15.65
S	2.37		2.06
$\hat{S}$	2.49		2.17
S_M	.79		.69

*I have numbered the subjects this way because it will help to make a point clear in a later section.

These data show that there was a difference of one point in the average rating. The question to be asked is whether this difference is significant. The route to the answer is one that you have traveled before. We ask, with the aid of an appropriate test of significance, what is the probability of obtaining a difference of 1.0 if the true difference is zero? This test requires that we estimate the standard deviation of the sampling distribution of differences, the *standard error of the difference* (S_{diff}) and then that we evaluate the difference of 1.0 in terms of the standard error of the difference. I show the calculations in the computational appendix. It turns out that the difference is far from significant.[4]

Tasters as a Random Variable, Within-Tasters Version

If you look at the table of data for the results just described, you will note that there is considerable difference of opinion among tasters as to the quality of the wine, and it might occur to you that this variability could make it difficult to obtain an effect of breathing even if it existed with the experimental method used in the first example. This is an important thought. Having had it, you can take the next step and ask: How could I reduce this taster-to-taster variation? And answer: perhaps by having the same tasters in both groups. This idea leads to the development of a within-subjects experiment.

Such an experiment would be a little more complicated to do than the between-groups version of the study because there is a new problem to consider. This is the problem of the order in which the tasters sampled the wines. Some would have to taste the freshly opened wine first, others would have to taste it first after it had a chance to breathe. It would be wrong, for example, to open the bottle, let people taste and do their ratings, wait for a standard amount of time, taste again, and then do the second rating. There would be two things wrong with such a procedure: (1) if the tasters knew about the effects of breathing on the palatability of wine, these expectations (demand characteristics, p. 79) would influence their ratings; (2) even if they did not have this knowledge, the effects of order of tasting and of breathing would be, as we say, confounded. Any differences obtained could be as sensibly attributed to the order of testing as to the effects of breathing.

Such problems, once recognized, are easy to solve, however. Suppose now that the experiment had been done properly. Ten individuals with the same qualifications as the 20 in our first imaginary experiment are selected. Each of them tastes the same wine as in the previous version, once immediately after it is opened, and once after breathing for a standard amount of time. Suppose that the data came out as shown in Table 7–5, which, you will recall, contains the same numbers as were obtained in the between-groups version of the experiment. This time, however, there are two ratings for each taster *and also a difference between these ratings for each taster*. This fact makes it possible to evaluate the results by means of a

[4]*Note to the bewildered.* The materials in this section and in a section or two in Chapter 8 get a bit technical. Please don't give up. If the details are too much, skim the arithmetic treatment and take my word for it that the conclusions are correct. I predict that you will find Chapter 9 worth the minor struggle that it takes to get to it.

slightly different application of normal-curve statistics. Consider each *difference* as a score for each subject and then carry out the following calculations: (1) Obtain the standard deviation of these scores. The table shows that it is 0.63. (2) Estimate the population standard deviation. Again the table gives the value $\hat{S}=.67$. Using the second of these values, estimate the standard error of the mean, shown in the table to be .21.

$$S_M=\frac{\hat{S}}{\sqrt{N}}=\frac{.67}{\sqrt{10}}=\frac{.67}{3.16}=.21$$

Note, however, that this is the standard error of a *mean difference*, S_{diff}. Thus it can be used in a Z test to evaluate the hypothesis that the true mean difference is zero, given that the obtained difference was 1.0.

$$Z=\frac{1.0-0}{.21}=4.76$$

If you look this value up in Table 6-7, you will find that a Z score this large will occur by chance very rarely. Thus you reject the hypothesis that the true difference is zero accept the hypothesis that the tasters detected the difference between the breathed and freshly opened wines.

The important thing to observe about this within-subjects result is that by comparison with the between-tasters version of this experiment, the

TABLE 7–5
Within-Subjects Wine-Tasting Experiment

Subject	Before Breathing	After Breathing	Difference
1	17.0	18.5	+1.5
2	17.5	18.0	+ .5
3	16.0	17.5	+1.5
4	16.5	17.0	+ .5
5	15.5	16.0	+ .5
6	15.0	15.5	+ .5
7	14.5	15.0	+ .5
8	13.0	14.0	+1.0
9	12.0	13.0	+1.0
10	9.5	12.0	+2.5
Mean	14.65	15.65	+1.0
S	2.37	2.06	.63
$\hat{S}$	2.49	2.17	.67
S_M	.79	.69	.21

results were very much more significant. This is because the within-subjects design removes the variance among individual tasters that were making it hard to detect the difference. You will be in a position to understand this better after you have read Chapters 8 and 10.

Wines as a Random Variable

Although the possibility would probably not occur to many people, the experiment *might* have been done another way. Locating *just one* of William James' people who could tell the difference between the top and bottom halves of a bottle of old claret, the experiment could have been done with a random sample of wines. Moreover, it could have been done using a between-wines design or a within-wines design.

Actually, the experiment would be a little tricky to do in either way—and expensive, because now it is going to take 20 bottles of wine for the single taster, whereas, allowing just a sip, the experiments with tasters as a random variable might have been done with just one bottle. In the between-wines version the procedure would be as follows: (1) Select a sample of 20 wines just as 20 individuals were chosen in the between-tasters experiment. (2) Divide these wines into two groups, one group destined to be allowed to breathe, the other not. (3) Have the single subject taste and rate the wines under two conditions. (4) Evaluate the difference in terms of S_{diff}. If the data were the same as in Table 7–4 (where the number of subjects would now be identifying numbers for each of the 20 wines), the difference of 1.0 point would not be significant.

The within-wines experiment would go as follows: (1) Select a random sample of 10 wines but get two bottles of each, one of each pair to be tasted freshly opened, one after breathing. (2) Have the single taster judge each of the wines under these two conditions, thus producing for each two ratings. (3) Evaluate the difference. Again if the numbers were those in Table 7–5, the result (effect of breathing) would be very significant.

The Generality of Experimental Results

The reason for describing this imaginary research in two different ways (with tasters and wines as the randomly represented variables) is that the two procedures lead to results with very different implications. When someone does an investigation, it is hoped that the results will reflect the truth of the real world. In this sense science is like drama. As Shakespeare says in *Hamlet*, its purpose is "to hold, as 'twere a mirror up to nature." It turns out, however, that the aspect of nature reflected in the mirror of scientific investigation depends upon the aspect sampled.

Another way to say this is that the population to which a result can be generalized depends upon the population from which random representatives were selected for the study. In the versions of the experiment with tasters as a randomly chosen variable, the results, *if significant*, mean that tasters in the general population of which the participants in the study are representative can detect the difference between the newly uncorked wine and the wine when it has been allowed to breathe—but only in the case of this one wine.

In the version with wines as a random variable, the justifiable interpretation is that the single taster can detect differences in the population of wines of which those selected are representative. But the results make the case only for that single taster, not for the population of tasters in general.

Toward Analysis of Variance

Obviously, the serious scientist of wine tasting would be dissatisfied with all of our sets of data because their generality is so limited. Such an individual would want to emerge from his investigation being able to generalize to both populations—the populations of tasters and of wines. In order to arrive at this strong position, he would be likely to set up an experiment in which he samples from both populations. He would select several tasters and have each of them taste several wines under both conditions. Such an experimental design would be a factorial design (p. 70) in which breathing or not was a manipulated variable and wines were a second selected independent variable. The method of analysis used to assess the outcome of such studies is called the *analysis of variance*. After a skirmish with the topic of correlation, for which it will take two chapters to introduce the basic ideas, I shall present the essentials of analysis of variance in Chapter 10.

SUMMARY–GLOSSARY

One way to view the content of this chapter is as a continuation of the hypothesis-testing materials presented in Chapter 6. The general argument in hypothesis testing always boils down to the following question: If the sampling distribution of the statistic obtained on some sample meets the criteria set forth in a set of assumptions, what is the probability of obtaining that statistic through the accidents of random sampling? The answer is a level of confidence with which the null hypothesis can be rejected. Actually, there is nothing very new in this presentation. The materials in this chapter have introduced you to some new technical concepts, but the line of reasoning is one with which you are already familiar. These new concepts, and some old ones, are those which follow.

Population. The entire universe of items of interest, frequently individual people.

Sample. A subset of items from a population.

Representative sample. A sample whose makeup matches that of the population, with respect to features related to phenomena being studied.

Biased sample. A sample that is not random.

Random sample. A sample in which every individual and every combination of individuals has an equal chance of being selected. The selection of one individual does not affect the probability of selecting any other. The selections are *independent*.

Statistic. A measure obtained on a sample.

Parameter. A measure for a population.

Symbols.

- M a sample mean (statistic)
- μ a population mean (parameter)
- S a sample standard deviation (statistic)
- σ a population standard deviation (parameter)
- $\hat{S}$ an estimate of S based on sample data: $\hat{S}=\sqrt{\frac{\Sigma d^2}{(N-1)}}$

Sampling distribution. The frequency distribution of a statistic that theoretically would be obtained if that statistic were obtained on an infinite number of samples.

Standard error. The standard deviation of the sampling distribution of a statistic.

More symbols.

S_M—the standard error of the mean. That is, the standard deviation of the sampling distribution of the mean.

S_{diff}—the standard error of the sampling distribution of the differences between means.

Confidence limits. The range within which the population value (parameter) of some measure falls with a stated degree of certainty. The example developed in this chapter involved the mean. Specified confidence limits for μ are the obtained sample mean plus or minus a range determined by an application of normal-curve statistics (Z score). Thus the 50% confidence limits for μ are $M \pm .6745 S_M$; the 95% confidence limits are $M \pm 1.96\ S_M$; the 99% confidence limits are $M \pm 2.58 S_M$.

Fiducial limits. Confidence limits. Fiducial derives from the Latin *fiducia,* trust. Thus confidence limits are the limits within which one can trust a statistic to fall—with a stated degree of certainty. Statistically speaking, trust is less than an absolute value.

Assumptions underlying statistical tests. The conditions that must be realized for valid application of a statistical test. The following assumptions apply to the tests described in this chapter.

1. The assumption of randomness, that the individuals assigned to groups for study are random samples.
2. The assumption of normality, that the sampling distribution of the statistic being evaluated is a normal distribution with a specific estimated standard deviation. This assumption is necessary to evaluate hypotheses in terms of areas under the normal curve.

Central limit theorem. Assuring the validity of the second assumption above, the distribution of totals (and, therefore means) of random samples drawn from any population is a normal distribution, provided only that the samples are large enough.

Between-subjects experiment. An experiment where different groups of subjects receive the different values of the independent variable.

Within-subjects experiment. An experiment where a single group of subjects receive all values of the independent variable.

Z test. A test used to evaluate hypotheses about the mean and about mean differences. The basic calculation involves the computation of a Z score, which is referred to in a table of areas under the normal curve.

Critical ratio. The Z test. It was called a critical ratio many years ago because a Z of 3.0 was taken as the criterion for rejecting the null hypothesis. This corresponds to a level of confidence of .0026, more stringent than is commonly used now.

t test. A test very much like the Z test except that the value of t is referred to a distribution that is slightly different from a normal distribution. More of this in Chapter 10 and the computational appendix.

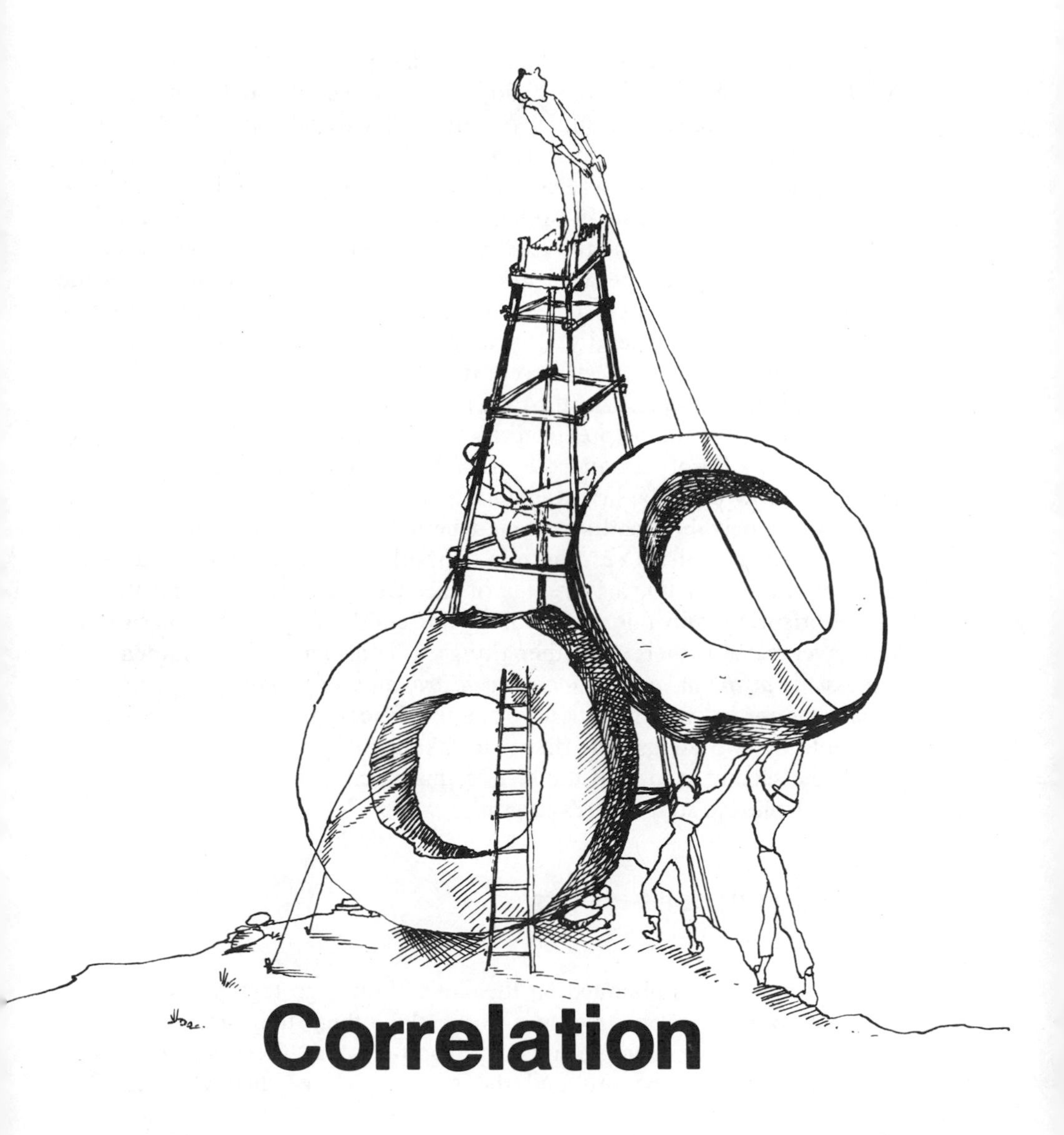

Correlation

Suppose that you are a young couple and that you have this problem: You would very much like to have a child, but you are concerned about the fact that there are already too many people in the world. Should you follow your personal desires and possibly add to the problem, or is the course of morality to put your own wishes aside for the common good? Finally, you resolve the conflict at least in principle. If in the long run the world will be a better world because of the existence of your offspring, you would be justified in having it; otherwise, not. What kind of person would leave the world a better place? Almost everyone knows the answer: a person who is trustworthy, helpful, friendly, courteous, kind, obedient, loyal, cheerful, thrifty, brave, clean, honest, reverent, and (for the sake of intellectual values and the example I want to use) intelligent.

But can such traits be predicted for the unborn? Does the fact that one of you has an IQ of 125 and the other has an IQ of 135 mean that your child will grow up to be bright, having perhaps your average IQ of 130? If you ask the neighborhood statistician he will tell you that it's a bit more complicated than that. Yes, there is a positive correlation of about +.60 between midparent IQ (the average of the IQs of the parents) and the IQs of the offspring, and this means that your child will probably be brighter than average. But there are other things to consider: the phenomenon of *regression to the mean*, the *sampling distribution of predicted IQs*, and the *standard error* in any prediction you make. Encouraged but confused, you decide to do two things. The first is to take the steps necessary to have the baby; the second is to read the rest of this chapter in order to remove the mystery from the statistician's jargon.

CORRELATION COEFFICIENTS

When two measures obtained on the same "thing" go together so that it is possible to predict one measure from the other, they are said to be correlated. The one example considered so far is that there is a correlation between the IQs of parents and the average IQs of their children. The "thing" in this example is the unit made up of two parents and one or more children. This means that it is possible to predict children's IQs from those of their parents. Parents with high IQs tend to have children with high IQs and parents with low IQs tend to have children with low IQs. More generally, the IQs of children tend to resemble those of their parents. The extent of this resemblance depends upon the size of the *correlation coefficient* describing the relationship.

A correlation coefficient is a number that shows the direction and degree of such relationships. There are actually many such coefficients, but by far the most commonly used is the Pearson product-moment correlation coefficient, called r. I shall present the entire discussion of correlation in this chapter in terms of r.[1]

This coefficient has values ranging from -1.0 through 0.0 to $+1.0$. The sign of the coefficient indicates the direction of the relationship (positive or negative). The absolute magnitude of the coefficient (sign ignored) indicates the strength or closeness of the relationship. A correlation of plus or minus 1.0 represents a perfect relationship. A correlation of 0.0 represents no relationship at all.

Positive and Negative Correlations

If high values on one of two correlated measures go with high values on the other, and low with low, the correlation is *positive*. Some examples:

1. As mentioned above, there is a positive correlation between parent's IQs and children's IQs.
2. There is even a stronger positive correlation between the IQs of identical twins; that is, the two members of a pair of twins usually have nearly the same IQs.
3. There is a positive correlation, state by state, between average income and the number of property crimes. Contrary to the predictions of some social scientists, who expect poverty and crime rate to go together, states with high income tend to have high rates of crime involving property.
4. There is a high positive correlation between the rated excellence of jug wines and their prices. In general, the more highly the wine is rated by expert tasters, the higher the price.

A *negative correlation* exists where high values on one measure go with low values on the other. Again, some examples:

1. For an informal illustration, consider an expression that my mother used to use with great frequency in her comments on something that I wanted to do that she didn't approve of: "The more I think of it, the less I think of it."

[1]On pages 244 to 245 of the computational appendix, I describe one other correlation coefficient, rho, which is based on ranks.

2. For a more formal example, think of what is often called a "speed–accuracy trade-off" in the performance of many tasks. The faster the person does the task, the lower his accuracy, and vice versa, the slower his speed, the greater his accuracy.
3. There is a negative correlation between the number of owner-occupied homes in a neighborhood and the number of large families. Large families tend to be poor families who cannot afford their own homes.
4. In a selection of English political districts, there is a negative correlation between the amount of open space devoted to parks and the proportion of accidents that are accidents to children. The more park space, the fewer the accidents to children.

High and Low Correlations

The absolute magnitude of the correlation coefficient indicates the closeness of the correlation. Some of the correlations mentioned above have the following coefficients: income and crime rate, +.50; qualities of wines and their prices, +.85; number of large families and number of owner-occupied homes, −.65; open space and accidents, −.86. Of these correlations the highest is the last one. It is the absolute size of the correlation that defines its magnitude.

Summary

The raw materials required to compute a correlation coefficient are two measures obtained on the same "thing." Often this thing is a single individual, but it can be other entities as well, for example, pairs of twins, jugs of wine, and political units. The correlation coefficient itself is a number between −1.0 and +1.0. The sign of the coefficient expresses the direction of the relationship. If high values on one measure go with high values on the other, and low with low, the correlation is positive. If high values on one measure go with low values on the other, and low with high, the correlation is negative. The magnitude of the correlation coefficient, direction ignored, reflects the extent of the tendencies just described. The meaning of this point will become clearer in the next section.

SCATTER PLOTS

One of the clearest ways to show the meaning of the concept of correlation is graphically, in the form of a *scatter plot*. In the construction of such graphs, the X axis represents one of the correlated measures. The Y axis represents the other.

Let us return to the example of the correlation between midparent IQ and children's IQs for an illustration. Suppose that our couple decides to check up on the statistician and, going to 10 other couples who have children, they collect the data shown in Table 8–1.

TABLE 8–1
Hypothetical Data on Midparent IQ and Children's IQ

Couple	*Midparent IQ*	*Average of Children's IQ*
1	125	110
2	120	105
3	110	93
4	105	125
5	105	120
6	95	105
7	95	75
8	90	95
9	80	90
10	75	80

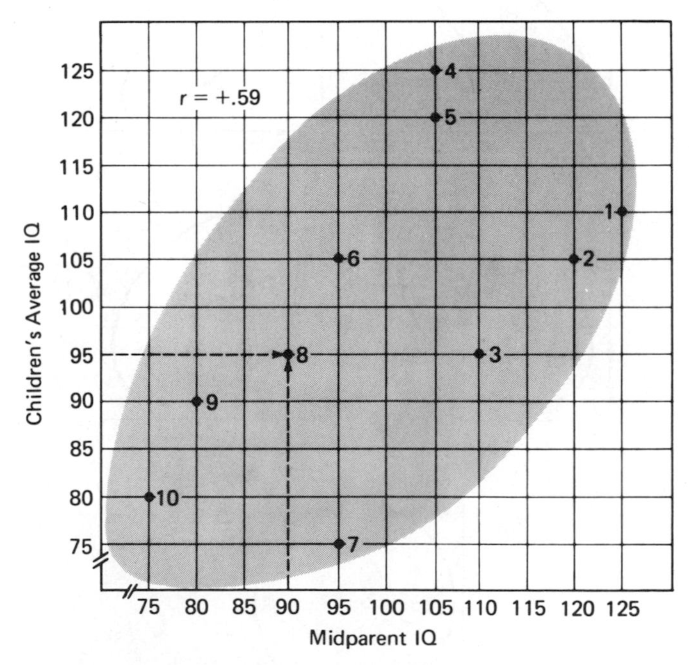

Figure 8–1 Scatter plot. These are the hypothetical data for the IQs of 10 pairs of parents and their children. Several points to notice: (1) midparent IQ is average IQ for parents; (2) plotting midparent IQ on the *X* axis and children's IQs on the *Y* axis is arbitrary; (3) the numbers near each point are the numbers from the "couple" column in the table of data (if these plots have any mystery for you, it would be a good idea to check a few); (4) the shape of the ellipse around the data depends upon the size of the correlation—it is fat if the correlation is low, thin if the correlation is high.

Just by looking you can see that there is some tendency for the IQs of children to resemble those of their parents, but the relationship is far from perfect. A scatter plot, of the sort presented in Figure 8–1, will provide considerable clarification. As a first order of business I would like briefly to go through the method of constructing such a plot.

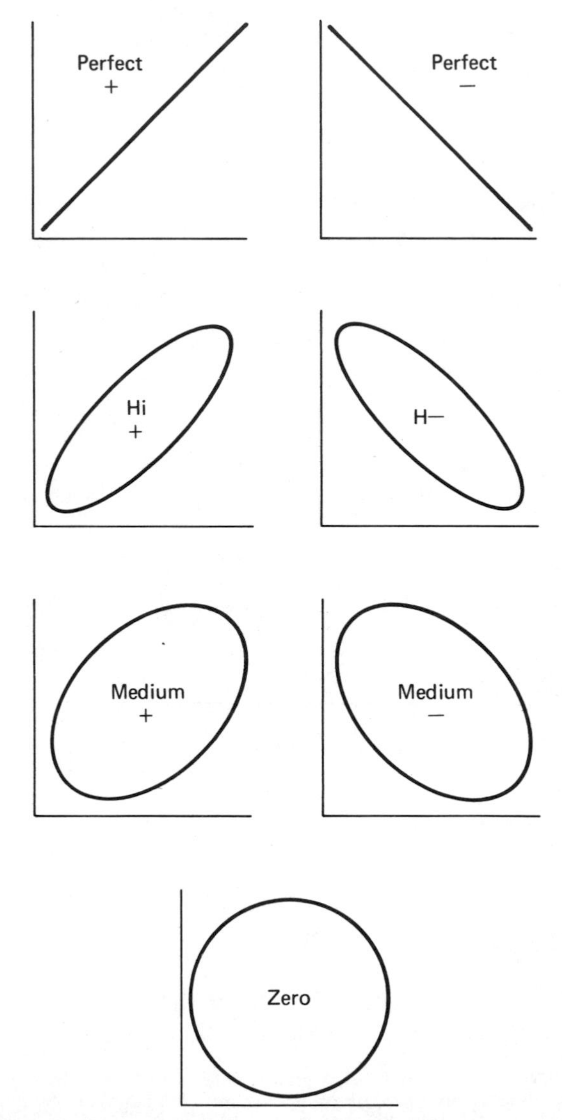

Figure 8–2 Schematic representation of the meanings of size and direction of correlations. Size (0 to ±1) affects the width of ellipse; direction (−1 to +1) is shown by the orientation of the ellipse.

These procedures are illustrated in Figure 8–1 for the data provided by couple 8. The point for this couple appears where a perpendicular line from their point on the X axis and a horizontal line from the average of their children's IQs would intersect. Although it is not standard practice, I have placed the identifying numbers of the couples next to each point so that you can refer to the table and check the procedure for plotting a few of the other points if it would be helpful.

The correlation coefficient represented by this scatter plot is +.59. In the next section I will show you where that number comes from. In the meantime, however, note that I have drawn in an ellipse that surrounds the points in the scatter plot. Its orientation from SW to NE shows that the correlation is positive but the shape of the ellipse shows that it is only moderately high. In general, the narrower the ellipse that will encompass the points in a scatter plot, the higher the correlation. At one extreme the "elipse" could be a circle, in which case the correlation would be zero. At the other extreme, the "elipse" could be a straight line, in which case the correlation is plus or minus 1.0. All of this is illustrated in Figure 8–2. Figures 8–3, 8–4, and 8–5 present the actual data which produced some of the correlations mentioned in the earlier discussion.

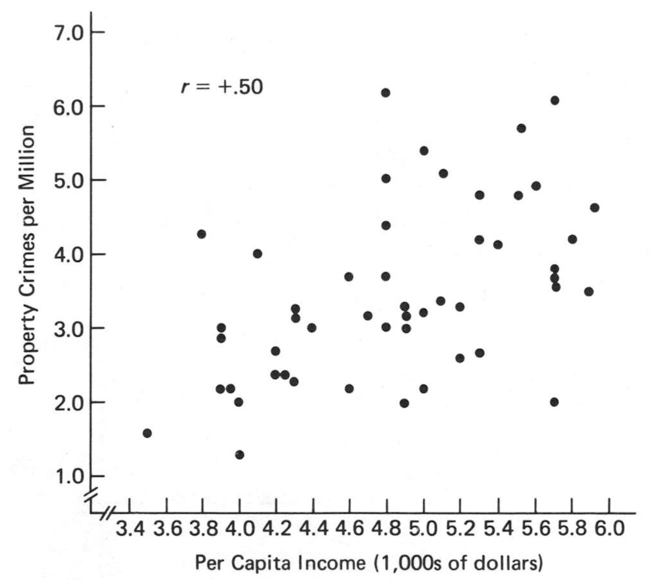

Figure 8–3 Scatter plot showing the relationship between income and crime rate. The measures are for each of the 50 states in the United States.

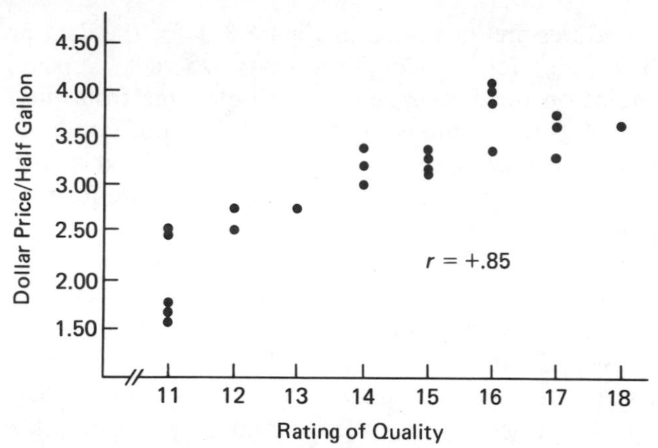

Figure 8–4 In jug wines you get what you pay for. Note that correlations allow predictions in either of two directions: (1) if you insist on a wine rated 15 or better, you will have to pay over $3; (2) if you spend no more than $3, you will get a wine rated 14 or less. The data were published in the American Express magazine *Travel and Leisure* in 1976. I averaged east and west coast prices and converted all of them to half-gallon prices before computing r.

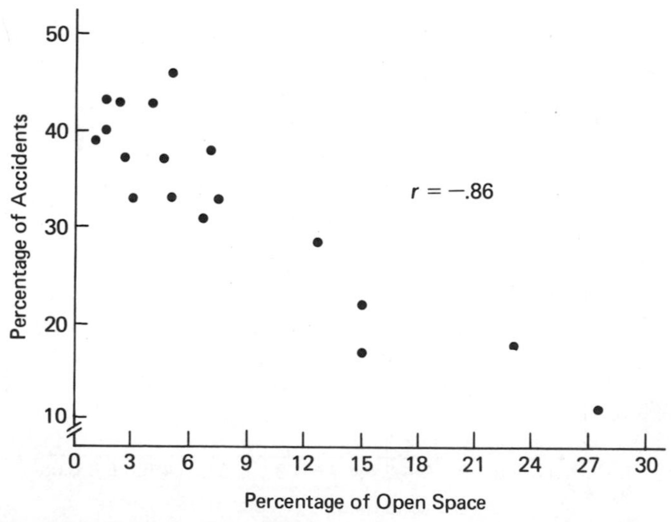

Figure 8–5 Scatter plot for a negative correlation. The data are from the book by M. J. Moroney, *Facts from Figures*, mentioned on page 106.

The Z-Score Formula

One formula for the Pearson product moment r is

$$r = \frac{\Sigma(Z_x \cdot Z_y)}{N}$$

where X is one of two measures to be correlated, Y is the second of these two measures, and Z, Σ, and N have their usual meanings. In words, this says that r is the "mean Z-score product." If one were to calculate r by this formula, the procedure would be to find the separate Z scores for all the X and Y measures for the units for which the correlation was to be calculated, to multiply these Z scores for each unit, and to determine the average. The procedure would be very tedious and, except for pedagogic purposes, it is hardly ever done.[2]

The tedious formula does have the useful virtue of making it easy to illustrate the calculation of a correlation coefficient. To do this, suppose that we correlate the data that were used to construct Figure 8–1. The set of midparent IQs and the IQs of their offspring appear again in Table 8–2, this time, however, with some other numbers required to compute the correlation coefficient.

As an incidental thing, you should note that I selected these numbers with some care so that they are roughly representative of the population of

TABLE 8–2
Computation of Correlation Between Midparent IQ and Children's IQ

Midparent		*Offspring*		
IQ	*Z Score*	*IQ*	*Z Score*	*Z-Score Product*
125	+1.63	110	+ .65	1.06
120	+1.30	105	+ .33	.43
110	+ .65	95	− .33	− .21
105	+ .33	125	+1.63	.54
105	+ .33	120	+1.30	.43
95	− .33	105	+ .33	− .11
95	− .33	75	−1.63	.54
90	− .65	95	− .33	.21
80	−1.30	90	− .65	.85
75	−1.63	80	−1.30	2.12
Mean 100	.00	100	.00	.586 = .59
S 15.33	1.00	15.33	1.00	r = .59

[2]See the computational appendix (p. 244) for a more common calculating formula. You will note that, although Z scores do not appear in this formula directly, the ingredients of Z are there.

IQs. The means of the two sets of IQs are both exactly 100, the known population mean. The standard deviations are both 15.33, again approximately the value of the population. Turning to the calculations, you will note that I have converted the IQ for midparents and offspring to Z scores. Then in the right-hand column I have multiplied the two related Z scores together and have obtained the mean of the 10 products. This mean of +.59 is the correlation coefficient. This procedure simply carries out the requirements of the formula.

$$r = \frac{\Sigma(Z_x \cdot Z_y)}{N} = \frac{(+1.63 \times +.65) + (+1.30 \times +.33) + \cdots + (-1.65 \times -1.30)}{10}$$

Scatter Plots Again

Another way to get at the concept of correlation begins with an important point about the significance of using Z scores rather than raw

TABLE 8–3
Correlation Between Prices and Ratings of Wines

Brand	*Rating (X)*	*Price (Y)*	Z_x	Z_y	$Z_x \cdot Z_y$
1	11	$ 1.57	−1.48	−2.11	+3.12
2	11	1.65	−1.48	−2.00	+2.96
3	11	1.75	1.75	−1.48	−2.75
4	11	2.49	−1.48	− .83	+1.23
5	11	2.49	−1.48	− .83	+1.23
6	12	2.51	−1.02	− .81	+ .83
7	13	2.70	− .57	− .54	+ .31
8	12	2.70	−1.02	− .54	+ .55
9	14	2.96	− .11	− .18	+ .02
10	15	3.15	+ .34	+ .08	+ .03
11	15	3.15	+ .34	+ .08	+ .03
12	14	3.18	− .11	+ .13	− .01
13	17	3.29	+1.25	+ .28	+ .35
14	15	3.32	+ .34	+ .32	+ .11
15	15	3.37	+ .34	+ .39	+ .13
16	16	3.37	+ .80	+ .39	+ .31
17	14	3.39	− .11	+ .42	− .05
18	17	3.57	+1.25	+ .67	+ .84
19	18	3.59	+1.70	+ .69	+1.17
20	17	3.69	+1.25	+ .83	+1.04
21	16	3.89	+ .80	+1.11	+ .89
22	16	3.99	+ .80	+1.25	+1.00
23	16	4.04	+ .80	+1.32	+1.06
24	15	4.26	+ .34	+1.63	+ .55
Mean	14.25	3.09	.00	.00	.85
S	2.20	.72	1.00	1.00	

scores in the calculation. What this does is to put both measures on the same scale, a scale with a mean of zero and a standard deviation of 1.00. This is true even when the original measures are very different. Take, for example, the set of data in Table 8–3 for the prices of jug wines and their ratings, the same data as were plotted in Figure 8–5. As a sort of review I have included the information required to do a correlation using the *Z*-score formula. In this case, $r = +.85$.

Figure 8–6 is a plot of price against rating in *Z*-score terms. The fact that there is a positive correlation between the two variables is obvious. A more important point will come through if you will notice that the dashed lines in the graph, horizontal and perpendicular to zero on the *X* and *Y* axes, divide the scatter plot into four quadrants. Reading clockwise from the upper left-hand quadrant, they are $-+$, $++$, $+-$, and $--$, in which $-$ and $+$ indicate whether the ratings and prices are above or below the mean. Now recall that $r = \Sigma(Z_x \cdot Z_y)/N$. If you examine the locations of the individual points in Figure 8–6, you will see that in this case the majority of them (21 of 24) are in the $--$ and $++$ quadrants. Since $(-)\times(-)$ and $(+)\times(+)$ both yield positive products, the *Z*-score

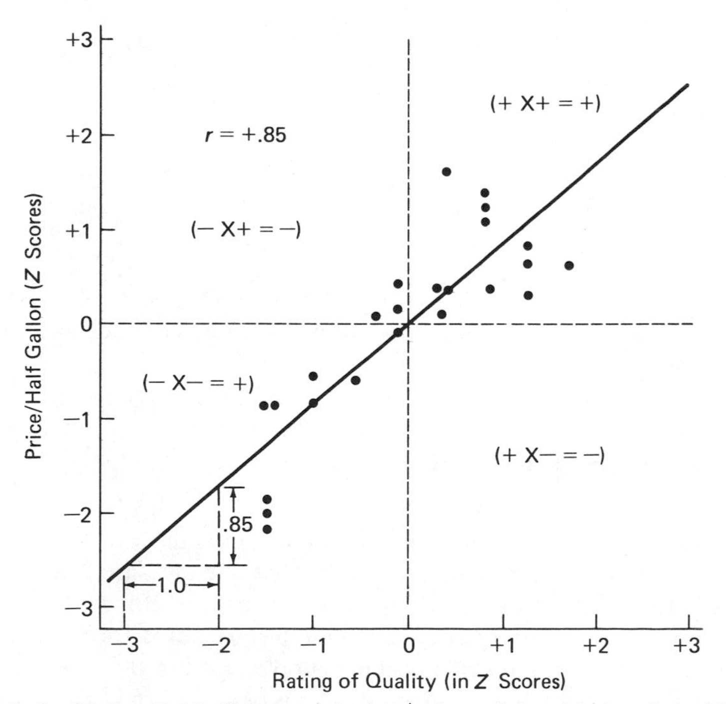

Figure 8–6 Scatter plot in *Z* scores of price/rating relationship for wines. Things to note: (1) the formula $r = \Sigma(Z_x Z_y)/N$ will yield mainly positive products where the correlation is positive, and (2) in *Z* score terms, *r* is the slope of the best-fitting straight line through the plotted points.

formula for r will give a result that is positive (e.g., $r = +.85$). This is because $\Sigma(Z_x \cdot Z_y)$ will be made up mostly of positive products from which only a few negative products will be subtracted before dividing by N to arrive at r.

Figure 8–7 is a scatter plot, in Z-score terms, for the negative correlation between open space and accidents. The point to note there is that the majority of the points are in the $-+$ and $+-$ quadrants, where multiplication produces negative products. This means that the calculations involved in the Z-score formula for r will produce a negative answer (e.g., $r = -.86$).

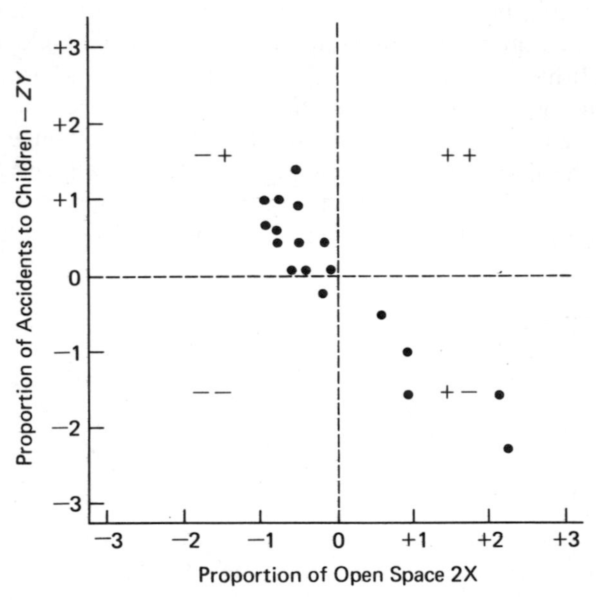

Figure 8–7 Scatter plot in Z score terms for the negative correlation between open space and accident rate to children.

REGRESSION AND PREDICTION

The great usefulness of the information contained in a correlation coefficient is that it makes it possible to predict one measure from another: children's IQs from parent's IQs, the price of a wine from a rating of its quality, or the relative frequency of children's accidents from the amount of open space, to mention some of the examples used in this chapter so far. The equation for predicting Y from X in Z-score terms is:

$$Z_y = r \cdot Z_x$$

One thing we can do with this *regression equation* is to begin to answer the question our young couple had about the IQ of their unborn child. Recall first that their average IQ was 130. Since the standard deviation of the IQ distribution is 15, this IQ corresponds to a Z score of 2.0.

$$Z_{\text{IQ }130} = \frac{130-100}{15} = 2.0$$

Recall second that the correlation between midparent IQ(x) and children's IQ(y) is +.60. Thus,

$$Z_y = +.60 \times 2 = +1.2$$

In terms of IQs, this means that the child's predicted IQ is 1.2 standard deviations above the mean IQ ($1.2 \times 15 = 18$ IQ points above the mean), or 118.

This example should also help you to understand an important aspect of predictions made with the aid of a correlation. Unless the correlation is ± 1.0, the predicted score is closer to the mean than the predicted-from score. This phenomenon, called *regression to the mean*, is easy to understand in terms of the equation, $Z_y = r \cdot (Z_x)$. Since r is never greater than 1.0, the predicted Z_y is always a decimal fraction (r) of Z_x. The extent to which regression to the mean occurs depends upon the size of the correlation. If $r = 1.0$, there is no regression at all. The predicted Z_y is exactly the same as the predicted from Z_x. If $r = 0.0$, regression is complete. The predicted Z_y is always 0.0, that is, the mean. As we saw earlier, the mean is always the best prediction to make in the absence of additional information because the errors of prediction will add to zero.

ACCOUNTING FOR VARIANCE

It is possible to develop quite a different type of understanding of the topic of correlation by approaching it from a different direction. The understanding to which this discussion is leading is that the existence of a correlation between two variables makes it possible to divide the variance in either of them into two components. One component is often called *variance accounted for*. This turns out to be the square of the standard deviation of the predicted values. The other component is often called *variance unaccounted for*. It turns out to be the square of the standard deviation of the errors of prediction, a term that I shall introduce later as the standard error of the estimate.

In order to present these ideas in terms that are more concrete, I will use once more the hypothetical data on midparent IQs and the IQs of offspring. The basic data appear once more in the first and fifth columns of Table 8–4. The discussion to follow not only makes the new points just mentioned but also provides a review of much of what has been presented earlier in this chapter.

TABLE 8–4
Predictions and Error in Prediction from Correlations

	Midparent IQ = x	Z_x	Predicted Z_y	Predicted IQ = Y	Actual IQ	Z_y	Error	Error in Z Scores
	125	+1.63	+ .96	114.72	110	+ .65	+ 4.72	+ .31
	120	+1.30	+ .77	111.80	105	+ .33	+ 6.80	+ .44
	110	+ .65	+ .38	105.83	95	− .33	+10.83	+ .71
	105	+ .33	+ .20	103.07	125	+1.63	−21.93	−1.43
	105	+ .33	+ .20	103.07	120	+1.30	−16.93	−1.10
	95	− .33	− .20	96.93	105	+ .33	− 8.07	− .53
	95	− .33	− .20	96.93	75	−1.63	+21.93	+1.43
	90	− .65	− .38	94.17	95	− .33	− .83	− .05
	80	−1.30	− .77	88.20	90	− .65	− 1.80	− .12
	75	−1.63	− .96	85.28	80	−1.30	+ 5.28	+ .34
Mean	100	.00	.00	100.00	100	.00	.00	.00
S^2	235	1.00	0.35	81.75	235	1.00	153.12	.65
S	15.33	1.00	.59	9.04	15.33	1.00	12.37	.81

Predicting Y from X

The formula for predicting Y from X in Z-score terms is $Z_y = r \cdot Z_x$. The first step in making a prediction on the basis of a correlation thus is to convert each predicted-from score to a Z score. The second column shows these Z scores, obtained from the formula

$$Z = \frac{X - 100}{15.33}$$

The mean and standard deviations appearing at the bottom of this column will serve as a reminder that the Z score distribution is the *unit normal curve*, with a mean of zero and a standard deviation of 1.00.

The third column of Table 8–4 lists the predicted Ys in Z score terms. The entries are simply .59 times Z_x. That is, this column shows the result of applying the formula $Z_y = r \cdot Z_x$. A point to note first is that the predicted Z_y is always less extreme than the predicted-from Z_x. This is the phenomenon of *regression to the mean*.

The summarizing materials at the bottom of the third column are very informative. Note first that the standard deviation of these predicted Z scores is .59, exactly the value of the correlation between X and Y. This is not a coincidence. Earlier, in presenting the concept of standard deviation I made the point (p. 111) that multiplying all the scores in any distribution by a constant multiplies the standard deviation by that same constant. The standard deviation of the values of Z_x was 1.0. Thus multiplying each Z_x by .59 would yield a new distribution with a standard deviation of exactly .59

For the major purposes of this discussion, a more important number to notice is the variance of the predicted values of Z_y, .35. This value needs to be seen in two ways: (1) it is the square of the standard deviation of these scores as a variance always is; and (2) in this case, in addition, it is r^2 ($.59^2 = .35$), since r is the standard deviation of these scores. This value has a very special meaning. It is the *proportion of variance in the Y scores accounted for* by the correlation between X and Y. Another way to say it is that the proportion of variance in Y,[3] accounted for by a correlation between X and Y, is r^2.

The fourth column in Table 8–4 presents materials that will help make this last point clearer. The entries are the average children's IQs predicted for each pair of parents. To review the procedure, the predicted children's IQ for the first midparent is +.96, an IQ .96 standard deviation above the mean. In terms of IQ points, where $S = 15.33$, this means that the predicted IQ is $.96 \times 15.33 = 14.72$ points above the mean, or 114.72. Accuracy to two decimal points is necessary to make later calculations come out close enough to be convincing.

You will note at the bottom of column 4 that the variance of these scores is 81.75. Now recall that the total variance in Y is 235.00. Evidently, the proportion of variance in Y accounted for by the correlation will be $81.75 \div 235.00 = .35$, the same value as r^2.

Errors of Prediction. An inspection of Table 8–4 will show you that the predicted IQ and the actual one were never exactly the same in this sample. The extent of this error appears in column 7 (4.72, 6.80, . . . , 5.28). Again the values at the bottom of the column are the important ones. The standard deviation of these entries (12.37) is the *standard error of the estimate* or the standard error of predicting Y from X, which I shall symbolize as $S_{y \cdot x}$.[4] The square of this term, $S^2_{y \cdot x}$, is a variance, the *variance in Y unaccounted for* by the correlation of +.59 between X and Y.

[3]In order to keep this presentation as straightforward as possible, I have developed the argument in terms of accounting for the variance in Y. Correlations are two-way streets, however. The whole argument can be turned around and one could predict midparent IQ from the average of children's IQs.

[4]This symbol does not imply multiplication. The expression can be read the standard deviation (S) of the Y values, given ($\cdot$) a value of X.

This variance unaccounted for has a value of 153.12. There are several things to notice about it. First in terms of proportion of total variance, the variance unaccounted for is $153.12 \div 235.00 = .65$. Since we saw earlier that .35 of the variance was accounted for by the correlation, this makes sense. But this making of sense implies something of considerable significance, that variances accounted for and unaccounted for are additive. A sort of check on this conception is provided by adding the variance in Y accounted for and unaccounted for: $81.75 + 153.12 = 234.87$. This differs from the expected total variance of 235 only because of errors in rounding.

A final point to notice is that converting the errors of predictions to Z scores yields, in the last column, a set of values (e.g., $4.72 \div 15.33 = .31$) whose variance is .65 and whose standard deviation is .81. The first of these is the proportion of variance unaccounted for. Since the proportion of variance accounted for (r^2) has been shown to be .35, the proportion of variance unaccounted for should be $1 - r^2 = .65$, as it is. The general thing to remember is that the proportion of variance unaccounted for equals $1 - r^2$.

Since $1 - r^2$ is a variance, its square root is a standard deviation, in this case $\sqrt{1 - r^2} = \sqrt{.65} = .81$. This value can be converted back into raw-score terms by multiplying by the standard deviation of these scores, 15.33. The more exact value of $\sqrt{.65}$ is .8062, and this times 15.33 is 12.3594, which again differs from the value of 12.37 calculated directly from errors because of the effects of rounding. This last example can be summarized quickly by presenting the formula for the standard error of the estimate:

$$S_{y \cdot x} = S\sqrt{1 - r^2}$$

Review

Obviously these materials on correlation hang together in a way that has a certain numerical elegance. They are also important enough to review in a somewhat more general way. To that end, suppose that we return for a last time to our young couple and their problem. Recall that the *parameters* assumed to exist in this example are the following: a mean IQ of 100 for midparents and children, a standard deviation of 15 for both populations, and a correlation of +.60 between the two variables. These values do not differ much from those used in the previous example. Assuming these particular ones simplifies the presentation a great deal, however.

Table 8–5 goes through the process of predicting children's IQs for a selection of midparents with IQs separated by 1 standard deviation.

Reviewing the steps in the procedures, they are as follows: (1) convert midparent IQ to Z scores; (2) multiply these Z scores by $r = +.60$ to get

TABLE 8-5

IQ	Z_x*	Z_y*$=r\cdot Z_x$	$Z_y\times 15$	$+100$
145	+3	+1.8	27	127
130	+2	+1.2	18	118
115	+1	+ .6	9	109
100	0	0.0	0	100
85	−1	− .6	−9	91
70	−2	−1.2	−18	82
50	−3	−1.8	−27	73

*Midparent IQ$=X$; predicted IQ for child$=Y$.

Z_y; (3) multiply Z_y by 15, the standard deviation of the IQ distribution, to turn Z_y into a deviation in IQ points; and (4) add 100, the mean of this IQ distribution, to this deviation to obtain the predicted IQ.[5] In passing you may want to note once more that the predictions, whether in terms of Z scores or IQs, show regression to the mean.

The standard error of the estimate in this example is 12.0 IQ points, obtained in the following way.

$$S_{y\cdot x}=S\sqrt{1-r^2}$$
$$=15\sqrt{1-(.60)^2}=15\sqrt{1-.36}$$
$$=15\sqrt{.64}=15(.8)$$
$$=12.0$$

All of this is summarized graphically in Figure 8–8. The solid line is a regression line connecting the predicted IQs for each predicted from IQ. The little normal distributions around the predictions are intended to suggest the idea that the actual IQs of the children would usually not be exactly the predicted value. Instead, they would form a normal distribution with the predicted Y as a mean and $S_{y\cdot x}$ as the standard deviation. Everything you know about normal distributions then applies to these little distributions. For our couple with the average IQ of 130, the average and most frequent IQ of an infinite number of offspring would be 118, but the chances are 68 in 100 that it will fall between 106 and 130 (the predicted IQ$\pm 1S_{y\cdot x}$). This particular point is illustrated by the distribution of IQs of children above a midparent IQ of 130 in Figure 8–8. Applying this line of reasoning in a couple of other cases, the chances are 50–50 that the IQ of a

[5]Sometimes things are a lot simpler if we resort to algebra. The formula for a Z score is $Z=(X-M)/S$. Thus $X-M$ is $Z\cdot S$ and $X=(Z\cdot S)+M$, which is all that is involved in these two last verbal steps.

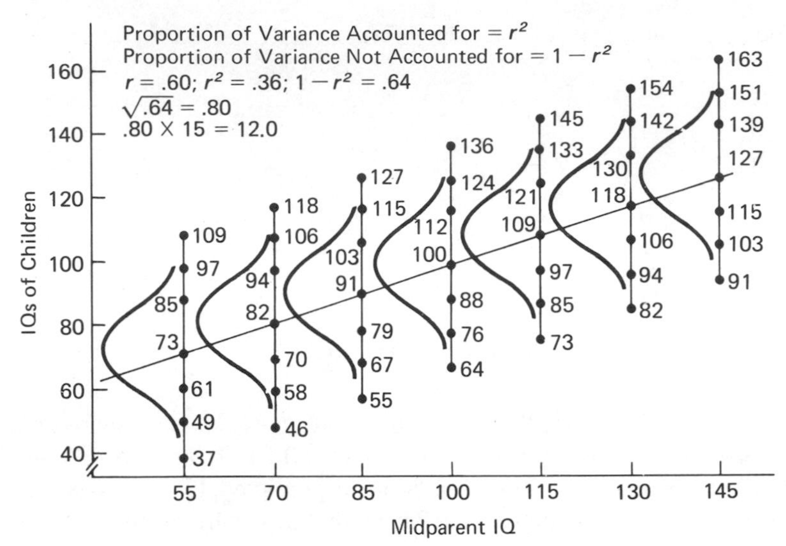

Figure 8–8 IQs of children predicted from midparent IQ. The heavy points are the predictions. The little distributions show the dispersion of actual children's IQs and thus are distributions of error (think of a correct prediction as one where the size of this error is zero). The standard deviation of each of these small distributions is 12.

child of parents with an average IQ of 130 will be between 110 and 126. This is the predicted IQ of $118 \pm .6745 \times 12$, the standard error of the estimate. The odds are approximately 95 in 100 that it will be between 94 and 142 ($118 \pm 2S_{y \cdot x}$).

SUMMARY–GLOSSARY

Things are correlated when they are "co-related" or go together. In the field of statistics, correlations express the degree to which two measures obtained on the same thing go together. This degree of relatedness is expressed numerically by a correlation coefficient.

Correlation coefficients put the individual who knows about them in a positive position to do two things: (1) to predict the value of either measure from a knowledge of the other, and (2) to state the amount of variation in one measure that is accounted for by variations in the other.

The prediction that one makes with the aid of a correlation coefficient is the average value of all the values of one variable that go with single values of the other. As this implies, there is a variation in the values of either variable that go with exact values of the other. The extent of this variation decreases as the strength of the correlation increases.

The variation just referred to is variance unaccounted for. Another way to say it is that this variance is the variance of the errors that one makes in predicting one measure from another. The averages predicted, seen in this context, represent variance accounted for. These last few statements summarize what it means to speak of a correlation as accounting for variance. The following important concepts take on meaning in this framework.

Correlation coefficient. Any of several different indices that express the degree, and in some cases the direction, of the relationship between two variables. The Pearson product-moment correlation coefficient, the only such measure discussed in this chapter, does both of these things.

r. The Pearson product-moment correlation coefficient. This measure has values ranging from +1.0 (perfect positive correlation) through .0 (no correlation) to −1.0 (perfect negative correlation.) The only formula for r presented in this chapter defines r as "the mean Z-score product," that is,

$$r = \frac{\Sigma(Z_x - Z_y)}{N}$$

Positive correlation. High values on one measure (x) go with high values on the other measure (y) and low values of x go with low values of y.

Negative correlation. High values of x go with low values of y, and vice versa.

Scatter plot. A graphic representation of the degree of correlation between two variables. Values of x go on to the abscissa; values of y go on the ordinate. The points *scattered* about this graph appear where axes extended from these coordinates for the values of x and y for a given thing would intersect.

Regression equation. The equation used to predict y from x, or vice versa. This chapter has presented everything in terms of predicting y from x, where the equation in Z-score terms is $Z_y = r \cdot Z_x$. To convert this to an actual predicted measure, it is necessary to do two things: (1) multiply the predicted Z_y by S_y, and (2) add the value thus obtained to the mean of y, which means subtract if Z_y is negative.

Variance accounted for. The variance of the values of y predicted from x (or vice versa) with the aid of the regression equation. In terms of proportions, this variance is r^2.

Variance unaccounted for. The variance in the errors of predicting y from x (or vice versa) with the aid of the regression equation. In terms of proportions, this variance is $1-r^2$.

Standard error of the estimate. The standard deviation of the population of y values predicted from a single value of x. This standard deviation is the square root of the variance unaccounted for, in terms of proportions $S_{y \cdot x} = \sqrt{1-r^2}$.

9 Uses and Misuses Of Correlation

Correlations, or something very much like them, play an important part in a great many socially significant issues. In this chapter, I will present a number of examples of the positive contributions of such materials as well as some of the hazards of interpretation.

"PERCENTAGE OF RELATIONSHIP"

My first example is a trivial one, involving little more than a fuzzy way of thinking. One sometimes hears a correlation interpreted as the "percentage of relationship" between two variables, an expression that you will now recognize as meaningless but as one that also might lull a listener into a feeling of false quantitive security. The only interpretation of that type that is justifiable is in terms of proportion of variance accounted for, which is r^2. An understanding of this point makes most of the correlations one is likely to encounter less impressive than they otherwise might seem. A correlation of $\pm.80$ accounts for 64% of the variance; one of $\pm.70$, 49% of the variance; and so on. Figure 9–1 is a graphic representation of the relationship between correlation and the proportion of variance accounted for.

CORRELATION VERSUS BASE-RATE ACTIVITY

Late in 1976 the federal government launched a mammoth campaign to inoculate the citizenship of the United States against swine flu. Before the end of the year, however, the program had, first, been interrupted and, later, aborted. The interruption occurred when about 100 elderly people died shortly after being inoculated. The termination of the program took place because approximately the same number of inoculated people developed the Guillain–Barré syndrome, an incapacitating, sometimes fatal,

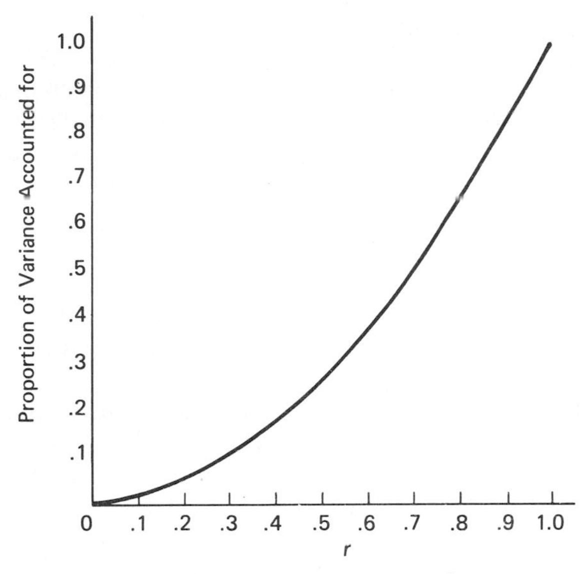

Figure 9–1 The value of r is not a percentage, although $r^2 \times 100$ is the percentage of variance accounted for.

paralysis. In both these cases the basis for action was a suspected correlation (and causal connection) between taking the flu shots and later lethal consequences.

The important question to ask in these two cases is whether the correlation actually exists. In the first case it quickly became clear that it probably did not. The death rate in aged people was about what it would have been if they had not been inoculated. The Guillain–Barré data are harder to evaluate. Various ways of looking at available statistics suggest that the incidence of the syndrome may be one to three cases per million in inoculated people and perhaps half or a third of that in uninoculated people. If so, there is probably a statistically dependable association between taking the flu shots and developing the syndrome.

The problem in these examples was that of deciding whether the incidence of death in aged people and the Guillain–Barré syndrome was greater in inoculated people than would be expected in terms of the naturally occurring *base rate* of these phenomena. In the first case, statistical evidence suggested that it was not. In the second case, it may have been. It is worth noting, in either case, however, that the probabilities involved were so low that it is far from clear that the increased risks were a sufficient basis for not getting the flu shots.

Correlation Is Not Causation

Even if the correlations discussed in the previous section existed, that fact would not prove that the flu shots were the *cause* of the bad consequences. It could be that, for some reason, people who decide to have the shots were particularly susceptible. Economic bracket, age, sex, race, and urban as opposed to rural residence are a few ways in which the inoculated and uninoculated might differ with such a result. For example, if the stresses of urban life favor death in old people (probable) or the development of the Guillain–Barré syndrome (questionable) and if more urban than rural people took the shots (probable), this could account for the association.

More generally, the fact that two events are correlated does not mean that one of them causes the other. There is, for example, a positive correlation between the number of storks' nests in Holland, year by year, and the birth rate in that country, but this does not prove the theory that storks bring babies. What it means is actually sort of the reverse. As the number of babies increases, for whatever reason, they need more houses to live in. Houses have chimneys and that is where storks build their nests.

The same point is even clearer in the next example. There is a high positive correlation between the number of fire engines in the several boroughs of New York City and the number of fires in those boroughs. Obviously, this means that the different amounts of fire-fighting equipment are required by different numbers of fires, not that fire engines start fires. In this last example, an obvious causal connection exists, but this observation leads to an important additional point. In and of themselves, correlations say nothing about the *direction* of causation.

A further consideration of the relationship between correlation and causation uncovers further complexities. In some cases the correlation has been produced by a third factor—the cause of both of the events that are correlated. For example, there is (or at one time there was) a negative correlation between the number of mules in our states and the number of Ph.D.s. The economic patterns of the states are responsible for both: heavily agricultural states, where mules are used, tend not to have as many educational institutions or industries of the type that hire doctoral-level personnel. In a similar vein there is a negative correlation, day by day, in New York City between the hardness of asphalt paving and the number of ice cream cones consumed, temperature being responsible for both. There is a positive correlation between the lengths of boys' trousers and the quality of their handwriting, age being the significant variable.

Turning to a less trivial example, there is a strong correlation between the amount of education people receive and their later incomes, but this does not mean that the amount of education causes a higher income. For one thing, more affluent people, who tend to send their children to college, also supply them with other advantages after graduation. For another thing, the connection may represent a selective process rather than a causal relationship. In order to get through school at any level a person has to clear a series of higher and higher intellectual hurdles. At some point those who cannot pass these tests leave the race. If the performances required to complete any level of education are like those required in economic competition later, those who survive are those best equipped for the real world outside the educational situation. This is to say that a large part of the contribution of education is not training but selection.

Finally, it is worth mentioning that the absence of a relationship, sometimes used to argue against a connection between two variables, can raise similar problems. One of the most time-honored arguments against capital punishment seems to have come from Samuel Johnson, who noted that hanging pickpockets apparently had no deterrent effect on the practice. Johnson's evidence was that pickpockets plied their trade with vigor at the public hangings of pickpockets. The trouble with this argument is that the number of pockets picked might actually have been fewer than at some other public ceremony of comparable size. If there were fewer, the common application of the argument would be in error.

In cases where correlation may actually mean causation, there is always a question of the directness of the causal connection. For example, recent seasons of drought in the West have called attention to the correlation between sunspots and dry weather. In this case there probably is a causal relationship involving the changes in solar winds that occur with sunspots. But a number of other phenomena are also correlated with sunspots: prices on the stock market, automobile production, the lengths of women's skirts, and the intensity of "Beatlemania," the popularity of the music of the Beatles. In all these cases, one can imagine that a remote causal set of connections exists. Drought means bad times for the farmers; they can't buy automobiles, so motor-car production goes down; this leads to lowered prices on the stock market and a change in national mood that affects people's preferences in skirts and music. All of this is speculation, of course. The causal relationships could be there, but the important point is that correlations do not prove causality and if the causal connection exists it occurs because of the operation of some third factor.

Partial Correlation

Sometimes the third variable responsible for a correlation can be canceled out statistically. Walter Kintsch once described the following example to make this point. There is a strong positive correlation between the number of bars in Montana towns and the number of churches. But this does not mean that the churches are there as places where the citizenry can go to ask forgiveness for the sins committed in the bars. Nor does it mean that the bars are there as places where people can go to forget the eternal damnation with which they are threatened once a week in church. All it means is that big towns have many bars and many churches and small towns have fewer of each.

There are fancier ways, called *partial correlation coefficients*, to remove the effect of the size of the community, but an easy way to see it is to imagine turning the number of bars and churches into rate measures—so many bars and so many churches per thousand of population. If that is done, the correlation between these rates is essentially zero.

One example in the previous chapter provides another example of such treatment of data. This is the correlation of $-.86$ between proportions of total accidents that were accidents to children and amount of open space devoted to playgrounds. Expressing the accident rate to children as a proportion of total accidents removes what, otherwise, would be an important third factor, population density, increasing the correlation. Places with little open space are heavily populated. Heavy population increases the number of accidents to everyone, including children. Showing that there is a negative correlation between proportion of accidents to children and amount of open space tends to suggest that the provision of space in which to play is a causal factor in reducing accidents to children.

Exactly the same situation exists with respect to traffic deaths on national holidays. Year by year the American Automobile Association comes forth with its dire predictions of the increased number of people who will die in traffic accidents during such holidays. The predicted increases occur more often than not. One might conclude on this basis that the chances of being killed in an auto wreck are enough greater over a holiday to provide a good reason for staying home. The conclusion does not follow, however. Some fraction of the increase—conceivably all of it—results from the greater number of drivers on the road. The rate of accidents could be the same.

THE CALCULATION OF "HERITABILITY"

In his famous study of "hereditary genius" carried out in the last century, Sir Francis Galton reported that very high talent runs in families. Galton's procedure was to identify eminent Englishmen—scientists, statesmen, judges, poets, and the like—and then to search for other eminent people among the relatives of these individuals. Not only did he find such eminent relatives, he also found that the percentage of them varied directly with closeness of the blood relationship. A portion of Galton's data appears in Table 9–1.

Newer studies of the correlations between the IQs of people with different degrees of blood relationships tell a similar story. Some data appear in Table 9–2. Obviously, the more closely people are related, the more similar their IQs are apt to be.

TABLE 9–1
Blood Relationship and Eminent Relatives

Nature of Relationship	Percentage of Eminent Relatives
Father–son	40
Brother	41
Uncle–nephew	20
Grandfather–grandson	15

TABLE 9–2
Blood Relationship and Correlations Between IQs

Nature of Relationship	r
Identical (MZ) twins	.90
Fraternal (DZ) twins (same sex)	.65
Nontwin siblings	.50
Either parent and child	.50
Grandparent and child	.15

The first thing one thinks, seeing such data, is that IQ must be inherited. The second thing one thinks is that the first thought was too hasty. People who are closely related tend to share much more similar environments than people who are less closely related. Conceivably, it is entirely the environmental effect that accounts for the parallel between blood relationship and similarity of intelligence. For years this environmental interpretation dominated the field.

More recent interpretations have assigned the determination of IQ partly to environmental factors and partly to heredity. Seen in the correct light, these interpretations assign part of the *variance* in IQ to environmental factors and part to heredity. This statement will lead you to anticipate that correlation coefficients will be involved in the assignment.

In determining the fraction of variance to be assigned to each set of factors the critical comparison is between identical (MZ, monozygotic) twins, who develop from a single fertilized egg, and fraternal (DZ, dizygotic) twins, who develop from two different eggs. MZ twins have identical heredity. They share 100% of the same genes. DZ twins, on the average, share 50% of the same genes. Both sets of twins share about as similar environments as is possible, especially if the DZ twins are of the same sex.

On these bases it could be argued that the fraction of the variance in IQ that is the result of heredity would be some function of the difference between the correlations for identical and fraternal twins. Different investigators have proposed different calculations on somewhat different logical grounds. One of the easiest to understand and also one of the most widely accepted is that the proportion of variance accounted for by heredity, h^2, is given by this formula

$$h^2 = 2(r_{MZ} - r_{DZ})$$

Applying this formula to the data in Table 9–2, we get

$$h^2 = 2(.90 - .65) = 2(.25) = .50$$

As you can see, the value of h^2 is determined, as suggested above, by the difference between the correlations for MZ and DZ twins. The multiplication by 2 comes about (in ways that I shall not develop) from the fact that MZ twins share twice the number of same genes as DZ twins do. The value of $h^2 = .50$ is on the low side of a range of values calculated for h^2 in different studies.

RELIABILITY AND VALIDITY OF MEASUREMENT

One of the most important applications of the methods of correlation, particularly in the social sciences, is in the standardization of measures, chiefly tests. Any useful measure has two essential characteristics, reliability and validity. Correlation coefficients, called reliability and validity coefficients in this context, provide a measure of the extent to which tests have these properties.

Reliability

Reliability refers to the extent to which a test or other measure performs consistently. A regular ruler is an example of a reliable measure; the commonly mentioned rubber ruler is an example of an unreliable measure.

The most direct way to determine the reliability of any measuring instrument is to measure the same things with it twice and compare the results. Since correlations by definition are indices of the correspondence between two measures obtained on the same thing, they provide an easy and natural quantitative way of making the comparison. This will tell you that reliability coefficients have values between .0 and 1.0 and that the higher the coefficient, the more reliable the test.

Most psychological tests have reliabilities in the range .70–.90. From what you now know about correlation and prediction, it is possible to give such numbers both a general interpretation and a specific one. What is involved is another analysis of the accuracy of predictions based on correlations. In this case think of it as predicting a later score from an earlier one on some test. You will recall that the amount of error in such predictions is reflected in the standard error of the estimate, $S_{y \cdot x}$.

Table 9–3 presents the analyses in question for reliability coefficients of IQ tests, examining the analysis for hypothetical coefficients ranging from .00 to .999. Working out the meaning of such a coefficient will provide a useful review and also some instructive new information. The review involves the standard error of predicting Y from X in proportions of variance and actual variance. Take, for example, the row beginning with a reliability coefficient of .90, approximately the value for the best IQ tests. The second entry in the table is just a reminder that the determination of $S_{y \cdot x}$, the term we need here, involves the calculation of the square root of variance unaccounted for, $\sqrt{1-r^2}$. The resulting value in the next column (.44) provides the general interpretation mentioned earlier. Whenever the reliability of a measure is .90, the standard error of the estimate will be

TABLE 9–3
Reliability and Precision of Measurement

Reliability	$1-r^2$	$S_{y \cdot x}$ (Proportions)	$S_{y \cdot x}$ (IQs)	$PE_{y \cdot x}$ (IQs)
.00	1−.00	1.00	15.00	10.12
.50	1−.25	.87	12.99	8.76
.60	1−.36	.80	12.00	8.09
.70	1−.49	.71	10.71	7.22
.80	1−.64	.60	9.00	6.97
.90	1−.81	.44	6.54	4.41
.95	1−.90	.32	4.74	3.20
.98	1−.96	.20	3.00	2.02
.99	1−.98	.14	2.12	1.43
.995	.99	.10	1.50	1.01
.999	.998	.05	.75	.51

.44 times the standard deviation of the measures being predicted, as required by the formula

$$S_{y \cdot x} = S_y \sqrt{1 - r^2}$$

The next column in Table 9–3 shows, in the specific case of IQs, that the standard error of the estimate is 6.54, .44 times the standard deviation of the IQ distribution, 15. This value begins to give you a feeling for what a reliability coefficient of .90 means. Possibly even a better sense of this meaning comes from the probable error of the estimate, $PE_{y \cdot x}$, in the last column of the table. The probable error of a score is the distance above and below a first score where 50% of the second scores would fall. The value of the probable error for any normal distribution is $.6745S$, a point that you may want to check by reference to Table 6–7 (p. 125). Thus the probable error of an estimate is $.6745S_{y \cdot x}$.

For an IQ test with a reliability of .90, $PE_{y \cdot x}$ is 4.41 points. This is the statistical basis for the common statement that an IQ score has an error of about 5 points. This is also the reason that psychologists who administer tests are very reluctant to give out exact scores. Unfortunately, the idea that measurements have an associated error is not well understood by the general public. People tend to place great faith in exact numbers. This lays the foundation for all sorts of potentially harmful interpretations, such as "My Mary is smart and your Jimmy is stupid because Mary has an above-average IQ of 101 and Jimmy has a below-average IQ of 99."

One reason for including very high reliabilities in Table 9–3 was to show you how very far away from accurate measurement even the best of our tests are. If you check, you will see that it would take a test with a reliability of .995 to produce an IQ with a probable error of one point. Something of the sort would have to exist before the Mary–Jimmy comparison above would make statistical sense.

Validity

Validity refers to the extent to which a measure predicts something important about the object of measurement. The physical measurement of the width of a table is valid, not so much because it tells you how wide the table is, as because it tells you whether you can get it through the door you want to, seat eight people for dinner, or cover it with a cloth you got for a birthday present. Physical measurement is valid because it allows accurate predictions of this type.

Other measurement is valid to the extent that it makes such predictions possible. IQ scores, for example, predict school performance, the sort of job a person can succeed at, and whether, if neurotic, he can profit from psychotherapy. Again correlation coefficients provide an index of the accuracy of such predictions. IQs are most valid for the prediction of school grades, where the validity coefficient is about +.60. The correlation is as low as it is (accounting for only 36% of the variance in school grades) for three main reasons.

1. IQ tests as we have seen are not perfectly reliable. This limits the size of the validity coefficient, which can be no higher than the coefficient of reliability.

2. School grades are even less reliable. This is the most serious problem encountered in establishing the validity of tests. The performance to be predicted, often called the *criterion*, with which test scores are correlated is often difficult to measure. Performance on a job is the most notorious example. Whether an employee's work is good, mediocre, or bad is usually determined for purposes of making important decisions (promoting or firing) in ways that are highly subjective and very unreliable. The validity coefficient can be no higher than the reliability of the criterion.

3. School performance (or any other complex activity) depends upon more than one ability—work habits, motivation, and special skills, as well as intelligence. This point may suggest to you that it would be useful if methods existed for taking all these capacities into account in making a prediction. They do and are called methods of *multiple correlation*. Although a description of these correlations is far beyond the limits of this book, it may be a little informative to say this much. Multiple correlations take several variables into account at the same time. They determine the predictive values of these variables and then put them all together to make a prediction that is usually better than is possible on the basis of just one variable.

VALIDITY AND HUMAN JUSTICE

Suppose that you are an administrator considering whether to retain or dismiss a member of your staff. Suppose that you have a recommendation from the immediate supervisor of this individual, which reads in part, "On a scale of 1 to 10, I would put Mr. X exactly in the middle in terms of his performance since he has been here." Suppose, finally, that one of your administrative goals is to upgrade the quality of your staff and that you want to keep on only people who will do better than average in their work. On the basis of the supervisor's rating, do you retain Mr. X or fire him?

If you ask the statistician in your organization for help, his response might be something like this: "Well, I've been keeping track of your supervisors' ratings and their predictions of future performance. I can tell you that there is a correlation of about +.50 between the supervisors' ratings and your own ratings later. Unfortunately, a correlation of .50 is so low that I cannot be very confident of any prediction I make. My best guess is that Mr. X will be about an average employee, but I could be way off. He might be a great success or an awful failure. In any event, you must understand that your situation is hopeless. Any policy you follow over a period of time is bound to be unfair to someone some of the time, either to the individual involved or to the oganization. There is no way that I can offer foolproof advice. I guess you're on your own."

The statistician's pessimistic position can be given graphic meaning, as in Figure 9–2. As you inspect the figure, you will begin to see why the administrator is in a no-win situation. The graph is a pictorial suggestion of what a scatter plot of the administrator's ratings against supervisors' earlier ones might look like if the correlation between the two sets of ratings is +.50. The oval shows the approximate pattern that the points might form in such a plot.

There is an important sense in which these data would have to be imaginary in part. Presenting them as a scatter plot implies that everyone rated was kept on, and his later performance evaluated, no matter what the supervisor's ratings were like. Obviously this would be poor policy, if only

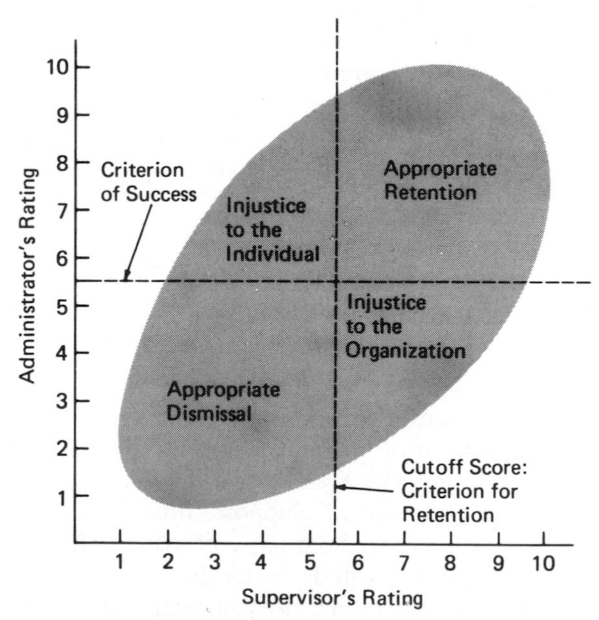

Figure 9–2 Justice and injustice brought on by imperfect validity. The thing to understand is that personnel decisions *always* reduce to a situation like this. The fact that they are based upon subjective impressions only makes things worse.

from the point of view of the supervisors' morale. The fact that such data are almost never available is the chief obstacle to arriving at satisfactory assessments of the validities of various kinds of information for personnel actions. There are also certain other consequences that I will get to in the next section. In the meantime it is important to think of the presentation in Figure 9–2 as showing the results of an ideal study that would hardly ever be carried out.

The dashed vertical line rising from a rating of 5.5 on the horizontal axis is graphic representation of the supervisor's judgment that Mr. X belongs exactly in the middle of a scale from 1 to 10 in terms of his performance since joining the organization. For purposes of developing the argument, suppose that at least an average performance is set as a criterion (or *cutoff score*) for retention in the organization. This would mean that everyone to the left of the line would be dismissed; everyone to the right stays on. The horizontal dashed line from a rating of 5.5 identifies the administrator's criterion of success. Everyone above the line is a success; everyone below it is a failure.

As you can see, these two lines cross in a way that divides the total population into four groups. Those in the upper right and lower left quadrants represent correct personnel decisions. The upper right quadrant contains the successes who rated highly enough to be retained. The lower left contains the low-rated people who would have been failures. The remaining two quadrants contain the tragedies in personnel action. Those in the upper left are the people who would have been successes but had ratings too low to be retained. In a sense the system has dealt them an injustice. The lower right quadrant contains the failures who were kept on because of their good ratings. They represent an injustice to the organization.

The first thing to note about this representation is that it is a realistic picture of the actual state of affairs that exists in situations requiring personnel decisions, whether or not there are formal ratings of employees. The graphic presentation is simply a way of making things more concrete. Since bases for prediction are always imperfect, there are always instances where competent people are let go and incompetent ones kept on, destined to become the "dead wood" in the organization.

The second thing to note about this representation is that the frequencies of these two types of mistake depend upon where cutoff scores for retention and criteria for success are set. Moving the cutoff score to the left —making retention easier and leaving the criterion of success where it is—produces more appropriate retentions, but it also keeps on more potential dead wood. Making retention more difficult—moving the cutoff score to the right—excludes more of the failures, but it increases the number of good people who are dismissed. If you try moving the cutoff score and the criterion of success around in various ways, you will find that nothing you can do is right. Whatever you do out of respect for human values hurts the organization, and vice versa.

Is there a way out of this dilemma? Not entirely because, as we have seen, measures of human abilities are so far from perfect. In general, however, the way to improve the situation is to increase the correlation between later performance and the measures used to predict it. The way to accomplish that, as in the case of the validity of IQ scores, is to increase the reliability of the predicted-from scores, and the criterion measures of later performance. This is likely to require increasing the amount and the variety of information used in both assessments.

Unfortunately, if many different measures are used for prediction and if detailed analyses of the criterion enter the picture, the process is likely to get pretty complex and that the *keep on–let go* decision might even have to be made by a computer. Many people find the prospect distasteful because it seems so impersonal. The fact of the matter is, however, that objective methods involving a quantitative treatment of data (sometimes called *actuarial prediction*) have repeatedly been shown to be more accurate than subjective methods involving an intuitive treatment of data (sometimes called *clinical prediction*). Looked at in terms of the consequences for the people about whom judgments are being made, the cold, actuarial methods turn out to be more humanitarian than the warm, clinical procedures that are in wider use.

PREDICTING FROM TRUNCATED DISTRIBUTIONS

Suppose that methods of predicting success in some enterprise attain a fair degree of validity, as the Scholastic Aptitude Tests (SAT) actually have in the case of success in college. What does the sensible person in charge of that enterprise do? Almost certainly he decides to make use of that information to do a better job of selecting people. Specifically this means setting a cutoff score and choosing for admission to his organization only people whose scores on the selection test are above the cutoff. Where he sets the score depends upon a number of factors, some of which I will discuss in a moment.

Suppose that the validity coefficient for a test is about +.60. A scatter plot for such a correlation might look something like the pattern in Figure 9–3, which is designed to show the effects of setting three different cutoff scores, labeled A, B, and C. These cut the horizontal axis of the graph at three points. In addition I have drawn in a horizontal line that defines the boundary line between success and failure.

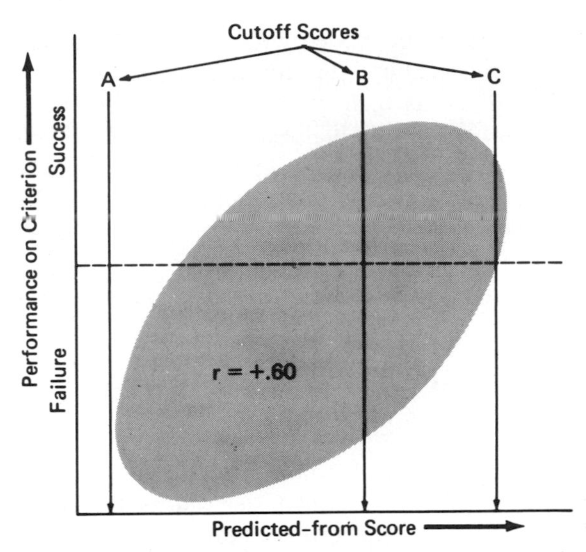

Figure 9–3 Predicting from truncated distributions. The use of higher and higher cutoff scores reduces *r* more and more. The thing to understand is that this does *not* mean that tests are no good, as has sometimes been alleged. The effect is a statistical artifact.

Turning now to the effects of the various cutoff scores, you will note that the one labeled A is not really a cutoff score at all. Everyone available is allowed into the organization. The correlation between the selecting measure and performance is +.60, and an inspection of the areas in the figure above and below the line dividing success and failure shows that more people fail than succeed.

Setting a cutoff score at the level indicated by the line labeled B does two things: (1) by excluding more potential failures than successes (to the left of line B), it improves selection: the people above the cutoff score are more often successes than failures; and (2) as you can probably recognize from the shape of the fraction of the original oval remaining, when scores below the cutoff are excluded, the correlation between the predicted-from score and performances on the criterion has been reduced. If you were to do the correlation for the people admitted to the organization, you might find it reduced from +.60 to +.30. The general point to understand is that anything that reduces the range of either of two correlated variables will reduce the correlation. In the language of statistics this is called the effect of doing correlations on *truncated distributions. Truncated* means "cut off" (as with a cutoff score) or "reduced in range." (Swimming "trunks" got their name in that way.)

The important thing to see about the reduction of correlations computed from truncated distributions of data is that it does not represent a failure of the tests involved. Quite the opposite: the tests are successful enough to be used in ways that appear to make them less useful.

Something like this has happened with the SAT tests mentioned earlier. For an unselected population, the correlation between SAT scores and college grades might be as high as +.60. As colleges become more and more selective, however, setting higher and higher cutoff scores for admission, the correlation has diminished. Things have now come to the point where the statistically unsophisticated presidents of some of our prestigious colleges are making public statements to the effect that the admissions tests are worthless. If the tests were to be abandoned for that reason, the error in such pronouncements would become apparent in only a year or two.

RISK/BENEFIT THINKING

One message that should come through from the materials presented so far in this chapter is that the real world, as well as the statistical world, is complicated. So many of the decisions that a person has to make in life have both good consequences and bad ones. All that happens as a function of the particular decision you make is that the balance of risks and benefits changes, but there is no way to make an important choice where the result is entirely positive.

In the case of taking shots for the swine flu (or undergoing almost any medical treatment), the benefit is to prevent one form of sickness, but this is at the risk of possibly developing another malady as a side effect of the treatment. Administrators trying to decide how to use information that is loosely correlated with the quality of performance on the job must weigh the inevitable injustices to workers on the one hand, and the organization on the other, that will result from any policy they develop. The more they emphasize the good of the organization, the more they do injustice to individuals. If, for example, they were to set their criteria of acceptance so high that all potential failures would be excluded in the process, they would also exclude many, many times than number of potential successes. Is such a policy ever justifiable? Ever even possible?

This leads us finally to criterion C in Figure 9–3. As you can see, it is set where only a few certain successes exceed the cutoff score. Suppose that I have been successful in setting it as I intended, and that only about .5% of the individuals tested have scores that exceed this cutoff. If you had to select 1,000 people who were sure to be successes, you would have to test 200,000 to get this number. Obviously, this would be a very costly program. Are there ever any circumstances that would warrant such an effort?

The answer appears to be "yes" for at least one case—if failure means death for the individual or other people. On this basis the American military establishment uses tests with fairly low predictive validity for the selection of candidates for pilot training in wartime. With an effectively limitless pool of candidates for pilot training available, it is possible to choose the very small fraction of them who are almost certain to succeed. Obviously, a substantial fraction of those excluded might have succeeded —but at the more frequent cost of thousands of dollars of wasted training money, millions of dollars of destroyed equipment, and the incalculable value of a human life.

NONMONOTONIC AND CURVILINEAR REGRESSION

There is evidence that the relationship between IQ and performance on certain jobs (such as those of taxi driver and checkout clerk in a grocery store) is *nonmonotonic*. That is, performance increases with increasing intelligence up to a point, following which there is a decrease. Beyond some optimal IQ, increasing intelligence interferes, possibly because the very bright are too easily distracted. If the relationship were perfect and regular, the points in Figure 9–4 would be what a scatter plot of the correlation would look like. The pattern formed by the points suggests that it would be possible to predict performance on the job perfectly from a knowledge of IQ. There is, in fact, a nonmonotonic correlation, *eta*, that would have a value of 1.0 in this case. The value of r computed for these

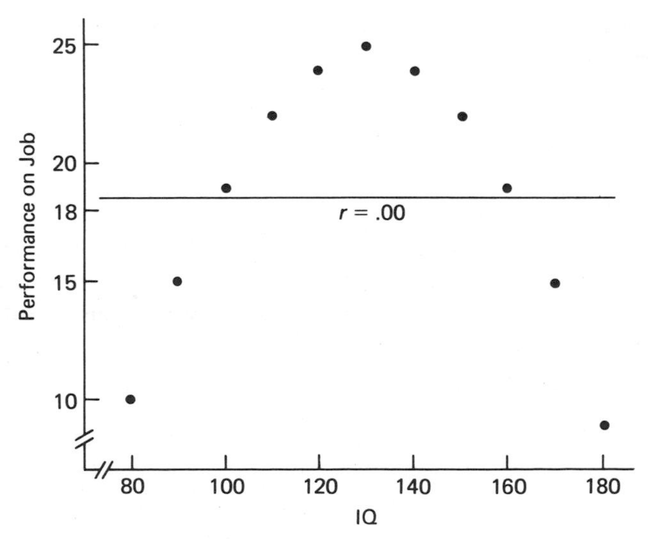

Figure 9–4 Perfect nonmonotonic correlation where $r=0$. An alternative correlation coefficient, eta, which is appropriate here has a value of 1.0.

data is .00, however. Using the regression equation based on this correlation, the prediction of performance would always be the mean, 18.64, no matter what the IQ.

The previous example is an extreme one. Figure 9–5 shows a case where the relationship between X and Y is a perfect monotonic one, but *curvilinear*. In this case the value of r is actually quite high, +.83, but it obviously underestimates the closeness of the relationship. It is particularly instructive to note how far the predicted values of Y are off when they are estimated on the basis of this correlation.

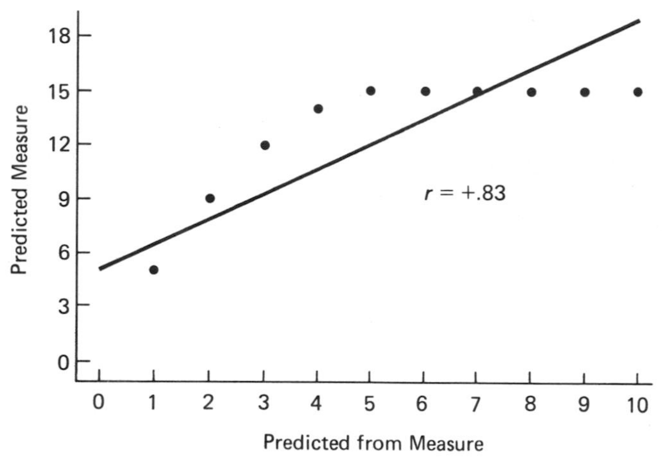

Figure 9–5 Perfect curvilinear correlation where $r = +.83$ and eta = 1.0. The thing to note here is that a correlation of .83, unusually high as actual data goes, can be obtained where this correlation would provide quite an inaccurate prediction.

The Fragility of *r*

One thing that this last example may do is to begin to give you a feeling of how far from perfect even a fairly high correlation is. We saw this point in one way in connection with our discussion of reliability. Still another way to make this point involves a numerical example. Consider the following pairs of X and Y scores.

X	1	2	3	4	5	6	7	8	9	10	11
Y	4	8	10	1	11	7	2	3	4	6	9

The correlation between these scores is as close to zero as I could make it without undue fiddling with the table, −.03. Now suppose, however, you change the first Y score to -40 or $+40$. If you do, r becomes $+.46$ in the first case and $-.52$ in the second. The point to get from this is that correlation coefficients, especially if they are based on small numbers of cases, are highly sensitive to the effects of a few extreme cases. Such correlations are very unreliable and should not be taken literally. One area of application where this consideration is very important is in the calculation of coefficients of heritability mentioned earlier. Since r is such a fragile number, it is to be expected that different data will lead to quite different estimates of the extent to which a trait is inherited.

Regression Errors

The phenomenon of *regression to the mean* was known to Francis Galton over 100 years ago. He called it "regression to mediocrity." Galton noted that tall parents tended to have tall children, but that the children were relatively less tall than the parents. More specifically, the child's height tended to be about half an inch above the mean height for every inch that a single parent's height was above the mean. Compared with the average height of both parents (midparent height), the child's height tended to be about four-fifths of an inch above the mean for every inch that midparent height was above the mean. As we have already seen, the *regression formula*, $Z_y = r \cdot Z_x$ demands the existence of this phenomenon, except for the rare cases where $r = \pm 1.0$.

The point I wish to make now is that an occasional error of interpretation assigns causal significance to the phenomenon of regression. Recall the example, with which the last chapter began, of the young couple with an average IQ of 130. The predicted IQ for one of their children was 118. Suppose that this couple has the child and that its IQ was exactly as predicted, 118. Conceivably they might feel somewhat responsible and guilty because their child's IQ was lower than either of theirs. They might think, for example, that there was something wrong with the way they brought up the child. Such thinking is an example of the *regression error*. It gives an unjustified causal explanation for a purely statistical phenomenon.

To take just one more example[1]—a true one—the instructors in a flight training school once adopted a policy of reward and punishment recommended by psychologists. They praised their trainees lavishly after each successful execution of a flight maneuver and punished them verbally after every poor performance. Contrary to their expectations based on psychological doctrine, praise seemed to lead to poorer performance on the

[1]This example is from D. Kahneman and A. Tversky, On the psychology of prediction, *Psychological Review*, **80**, 1973, pp. 237–251. The whole paper, which covers other misinterpretations of statistical data, is worth reading. If you have followed this book up to this point, with understanding, the level of presentation will be within your grasp.

trainee's next attempt at the maneuver. The punished trainees usually showed an improvement.

A possible interpretation of these outcomes might be that only punishment is effective in altering performance. This interpretation would be in error. The results are a regression artifact. Progress in flight training is slow and the correlation between successive performances is low. Following a particularly good performance regression to a poorer one is to be expected. As Kahneman and Tversky point out, the implications of these artifactual effects are enormous. We all live lives where our attempts to improve people's behavior by reward leads to our own punishment. Our attempts to produce improvements by punishment are rewarded with success. No wonder we have developed what has been called "the blame and shame society."

The phenomenon of regression to the mean is with us constantly. The children of very bright people are often disappointments. The best student on the first examination is not apt to be the best on the second, nor is the worst student apt to be the worst on two successive examinations. The company that performed most spectacularly on last year's stock market will probably not do so well this year. There is a widespread suspicion that the second year in the major leagues is unlucky for the best-performing rookies of the previous year. Every one of these examples provided an invitation to *explain* the changed performance. This is an invitation to commit a regression error. The invitation should be declined.

SUMMARY–GLOSSARY

Correlation coefficients merely express the degree and direction of relationships between variables. They say nothing about the kind of relationship. Most specifically, they say nothing about causation—neither that a causal relationship exists or that one does not exist. Interpretations of causality depend upon nonstatistical consideration, no doubt the most important of these being a question of the nature of causality itself. Since the philosophers have not yet solved this basic problem, it is obviously risky to make causal statements on the basis of a correlation.

Another risky business involves the using of correlation coefficients for the making of predictions, their main practical usefulness. As we have seen a number of times, a given correlation (r) only accounts for a proportion (r^2) of the total variance in the predicted variable. This means that predictions based on correlations can be subject to enormous degree of inaccuracy unless the correlation is very high.

The interpretation of correlation coefficients is also fraught with other, somewhat more specific, opportunities for making errors. A good many of these, as well as some of the useful properties of correlations, are suggested by the following concepts and their definitions.

Base rate. The natural frequency of some event. Base rates become a problem when they must be taken into account in concluding that a correlation exists between two variables.

Partial correlation. A type of correlation coefficient that tells the degree of correlation between (typically) two variables when the value of a third, related to both of them, is held constant.

Heritability. The proportion of variance in some trait *in some population* (very important) that is due to genetic causes.

Reliability. The extent to which a measure gives dependable values; the extent to which two measurements of the same things with the same measuring instrument rank these things in the same order.

Validity. The extent to which a measure predicts something of interest about the thing measured.

Criterion. In connection with validity, the property predicted by the measure.

Cutoff score. A score set as the minimum requirement for admission to a program, retention in it, or some other action where individuals must be selected on the basis of expected success.

Standard error of the estimate. The standard deviation of the distribution of predicted measures where the prediction is based on a correlation.

$$S_{y \cdot x} = S_y \sqrt{1 - r^2}$$

Actuarial prediction. Prediction on the basis of formulas and objective data, these days often with the aid of a computer.

Clinical prediction. Prediction on the basis of subjective impressions. In almost every study comparing clinical and actuarial prediction, the latter has been found to be superior.

Truncated distribution. A distribution with a restricted range (e.g., as a result of using a cutoff score). The effect of truncating a distribution is to lower the correlation with predicted measures.

Nonmonotonic regression. The curve relating mean values of Y (predictable from X) to X is a nonmonotonic function.

eta. A correlation coefficient that may be used where regression is curvilinear or nonmonotonic.

Curvilinear regression. The curve relating mean values of Y (predictable from X) to X is not a straight line. In such cases, r does not indicate the true degree of correlation because r assumes linear regression.

Regression to the mean. Unless $r = 1.0$, the regression equation always predicts a Z_y that is closer to the mean than Z_x, from which it was predicted.

Regression equation. In Z score terms, $Z_y = r \cdot Z_x$.

Regression error. Giving a causal interpretation to the phenomenon of regression to the mean.

ANOVA

There is good evidence that the stresses of life are responsible for a variety of problems of health. As you will see this statement means, among other things, that there must be ways to assess these stresses. There are, and work with human beings has led to the development of an inventory of some of our most important difficulties. This work is an application of what is called the method of *ratio scaling*. The death of a spouse is the greatest stress a person can experience. In the scaling procedure it receives an arbitrary value of 100. Other stresses can be judged by people as a subjective percentage of such stress. These other events are assigned these percentages as an index of their stress value. As it turns out, almost any change in a person's situation is stressful. Table 10–1 gives the stress units associated with various events in life. One thing to notice is that even "good" changes are stressful. I have marked with an asterisk some of these positive happenings that are, nevertheless, stressful. Psychosomatically speaking, the best things in life are not free.

One of the best studies of the effects of stress on health was done by Richard Rahe,[1] a psychiatrist who works for the Navy. Rahe has access to a large population of people, Navy personnel, who are fairly similar in background and about the same age. He obtained total stress units for almost 3,000 Navy men and then studied their reactions under the closely controlled conditions provided by a training cruise. The dependent variable in the study was the number of times the men answered "sick call" on this cruise. The independent variable was four different totals of stress units accumulated in the six months prior to the cruise. As a matter for review, you may want to note that this is a multivalent experiment and that the independent variable in this study is a selected independent variable. The method is what I have called quasi-experimental, employing unobtrusive measures.

The results of the study showed a relationship between stress level and frequency with which the men were sick. The four means were: very low stress, 1.4; low stress, 1.6; high stress, 1.7; and very high stress, 2.1. The question we have to face now is whether these differences are significant.

[1]R. H. Rahe, Subjects' recent life changes and their near-future illness reports, *Annals of Clinical Research*, **14**, 1972, pp. 250–265.

TABLE 10–1
Life-Change Events

	Stress Units
Family	
Death of spouse	100
Divorce	73
Marital separation	65
Death of close family member	63
*Marriage	50
*Marital reconciliation	45
Major change in health of family	44
*Pregnancy	40
*Addition of new family member	39
Major change in arguments with wife	35
Son or daughter leaving home	29
In-law troubles	29
Wife starting or ending work	26
Major change in family get-togethers	15
Person	
Detention in jail	63
Major personal injury or illness	53
Sexual difficulties	39
Death of a close friend	37
*Outstanding personal achievement	28
*Start or end of formal schooling	26
Major change in living conditions	25
Major revision of personal habits	24
Changing to a new school	20
Change in residence	20
Major change in recreation	19
Major change in church activities	19
Major change in sleeping habits	16
Major change in eating habits	15
*Vacation	13
*Christmas	12
Minor violations of the law	11
Work	
Being fired from work	47
Retirement from work	45
Major business adjustment	39
Changing to different line of work	36
*Major change in work responsibilities	29
Trouble with boss	23
Major change in working conditions	20
Financial	
Major change in financial state	38
Mortgage or loan over $10,000	31
Mortgage foreclosure	30
Mortgage or loan less than $10,000	17

ONE-WAY ANALYSIS OF VARIANCE

From what I have presented so far in this book, the only possibility for determining the significance of the differences among these means would be to use a Z test and compare the means of group 1 with groups 2, 3, and 4; group 2 with groups 3 and 4; and group 3 with group 4. This would be wrong, however, for a very straightforward reason. Suppose that, because of sampling errors, the mean of one of the samples (say group 3) happens to be much lower or much higher than the population mean. Because of this, you reject the null hypothesis in the first comparison between groups 1 and 3, committing thereby a Type I error. Since the mean for group 3 is involved in several of the other comparisons, the same error will occur in each of these comparisons, compounding a misinterpretation. Another way to put it is that the several comparisons are not independent of each other. It will be necessary to find another way to deal with these data. One way is with the analysis of variance, commonly abbreviated ANOVA.

The Logic of Analysis of Variance

Analysis of variance takes an approach to the problem which evaluates the significance of the differences among several means, all at the same time. The F test or F ratio used in these methods provides an answer to this question: *Is the variance of the means obtained in a study too great to permit the conclusion that they are all the means of random samples drawn from the same population*? The null hypothesis is that the variance among the means is not too great. If the null hypothesis can be rejected, this means that the independent variable in the experiment, in effect, has created different populations, from which the sample means were drawn. The methods of analysis of variance depend upon ideas that are closely analogous to ideas developed in Chapter 8.

Partitioning Variance

In Chapter 8 I showed that a knowledge of the correlation between two measures allows you to divide the variance in either of them into two components. One component was the variance of a set of means predicted by a regression equation. The other component was the variance of a distribution of scores that surrounds these predicted means. Suppose that we symbolize the total variance as S_T^2, the variance of the means as S_M^2 and the variance surrounding the means as S_{WG}^2 (within groups). This allows us to say that

$$S_T^2 = S_M^2 + S_{WG}^2$$

You may recognize the term S_M^2 in the equation as the square of the standard error of the mean and S_{WG}^2 as the square of a sample standard deviation (S^2). If you recall (p. 141) that

$$S_M = \frac{\hat{S}}{\sqrt{N}}$$

so that

$$S_M^2 = \frac{\hat{S}^2}{N} \quad \text{and} \quad NS_M^2 = \hat{S}^2$$

and that

$$\frac{N}{N-1}S^2 = \hat{S}^2$$

it is clear that both of these terms can give us an estimate of the population variance, $\hat{S}^2$. Such estimates are the basic ingredients of ANOVA. Analysis of variance asks, now in simple terms, whether these are probably estimates of the variance of the same population.

Partitioning Sums of Squares

Although the preceding equations above can be made to yield the estimates of population variance that go into an F ratio, they are a little cumbersome. The formula for an estimate of population variance that we have used before is this:

$$\hat{S}^2 = \frac{\Sigma d^2}{N-1}$$

A more direct way to get to the terms we need follows the same pattern. It begins with the calculation of a total sum of squares (Σd_T^2) and then goes on to break this total down into a sum of squares between groups (Σd_{BG}^2) and a sum of squares within groups (Σd_{WG}^2). Σd_{BG}^2 provides the estimate of population variance based on means. Σd_{WG}^2 provides the more traditional estimate based on sample variance. Once more the materials covered in connection with correlation are relevant.

In Chapter 8, I outlined the methods of predicting Y from X in the case of correlated data, with the aid of the regression equation $Z_y = r \cdot Z_x$. One example employed in that chapter predicted the IQ of a child born to parents with an average IQ of 145. Since this average IQ is 3 standard deviations above the mean, and since the correlation between midparent

IQs and children's IQs is +.60, the equation becomes $Z_y = +.60 \times 3 = 1.8$. In terms of IQs, this is 127, an IQ 1.8 standard deviations above the mean. The predicted IQ is an average IQ, a point about which there would be a normal distribution of IQs of individual children. One child born to such parents, whose IQ is 103, will serve to introduce the concepts actually required for the calculation of F in the analysis of variance.

This child is a member of a group whose mean IQ is 27 points above the population mean ($127-100=27$). On the other hand, the child's IQ is 24 points below the mean for his own group ($103-127=-24$). Thus the 3-point difference between the child's IQ and the population mean of 100 is made up a difference of +27 points *between* his group mean and the population mean plus a difference of −24 points of the child *within* his own group, and $27-24=3$.

As this example shows, the difference (d) between any individual score and the general mean in a set of scores divided into subgroups is the sum of a between-group difference (d_{BG}) between the subgroup mean and the general mean and a within-group difference (d_{WG}) by which the individual score falls above or below the mean within its own group.

$$d = d_{BG} + d_{WG}$$

With this equation before you it will probably come as no surprise that another equality also holds. This is that

$$\sum d_T^2 = \sum d_{BG}^2 + \sum d_{WG}^2$$

Life Stress and Illness: An Example

It is only a brief step from sums of squares ($\sum d^2$) to variance. Even so, it seems a good idea to return to the example of life stress and illness as an aid to taking it. Evaluated by the methods of analysis of variance, the increase in illness with increasing life stress was highly significant ($p < .01$). Rather than repeating the actual analysis used in the original study, however, I would like to present some highly simplified data that will serve to make the essential points. The data appear in Table 10–2.

The entries labeled X are hypothetical numbers of sick calls for 40 men, 10 in each of four stress-level groups. Means and variances for these groups and for the total appear at the bottom of the table. Within the body of the table, three deviation scores are entered for each score: d_T, which is the total deviation of each X from the mean ($X - M$), d_{BG}, the difference between each of the four group means (1.0, 2.0, 3.0, and 4.0) and the mean for the total (2.5) and d_{WG}, the difference between each score and the mean

TABLE 10–2
Number of Sick Calls by Individuals with Four Levels of Life Stress

	Very Low				Low				High				Very High				Total
	X	d_T	d_{BG}	d_{WG}	X	d_T	d_{BG}	d_{WG}	X	d_T	d_{BG}	d_{WG}	X	d_T	d_{BG}	d_{WG}	
	0	−2.5	−1.5	−1.0	1	−1.5	− .5	−1.0	2	− .5	+ .5	−1.0	2	− .5	+1.5	−2.0	
	0	−2.5	−1.5	−1.0	1	−1.5	− .5	−1.0	2	− .5	+ .5	−1.0	3	+ .5	+1.5	−1.0	
	0	−2.5	−1.5	−1.0	2	− .5	− .5	.0	2	− .5	+ .5	−1.0	3	+ .5	+1.5	−1.0	
	1	−1.5	−1.5	.0	2	− .5	− .5	.0	3	+ .5	+ .5	.0	4	+1.5	+1.5	.0	
	1	−1.5	−1.5	.0	2	− .5	− .5	.0	3	+ .5	+ .5	.0	4	+1.5	+1.5	.0	
	1	−1.5	−1.5	.0	2	− .5	− .5	.0	3	+ .5	+ .5	.0	4	+1.5	+1.5	.0	
	1	−1.5	−1.5	.0	2	− .5	− .5	.0	3	+ .5	+ .5	.0	4	+1.5	+1.5	.0	
	2	− .5	−1.5	+1.0	2	− .5	− .5	.0	4	+1.5	+ .5	+1.0	5	+2.5	+1.5	+1.0	
	2	− .5	−1.5	+1.0	3	+ .5	− .5	+1.0	4	+1.5	+ .5	+1.0	5	+2.5	+1.5	+1.0	
	2	− .5	−1.5	+1.0	3	+ .5	− .5	+1.0	4	+1.5	+ .5	+1.0	6	+3.5	+1.5	+2.0	
Mean	1.0				2.0				3.0				4.0				2.50
S^2	.60				.40				.60				1.20				1.95

of the group of which it is a member. If you check a few of the entries, you will find as mentioned earlier that $d_T = d_{BG} + d_{WG}$. With these data available, I can easily do two things: (1) demonstrate the truth of the two equations presented earlier in the chapter,

$$S_T^2 = S_M^2 + S_{WG}^2$$
$$\Sigma d_T^2 = \Sigma d_{BG}^2 + \Sigma d_{WG}^2$$

and (2) move on to the methods of analysis of variance.

Components of Variance. The total variance of the scores in Table 10–2 is 1.95. To see how this breaks down into two additive components, consider first the simple calculations in Table 10–3. These calculations treat each mean from Table 10–2 as a score and compute the variance of these scores by the formula

$$S^2 = \frac{\Sigma d^2}{N}$$

where S^2 is actually S_M^2 because the original scores are means. The value for this variance is 1.25.

The second fraction of variance is within-groups variance, the average of S^2 for the four separate groups from Table 10–2: $.6 + .4 + .6 + 1.2 = 2.8 \div 4 = .70$. Within-groups variance plus the variance of the means is $.70 + 1.25 = 1.95$, the calculated variance for the total group. This shows, as mentioned earlier, that

$$S_T^2 = S_M^2 + S_{WG}^2$$

About the only new point for you to see is that S_{WG}^2 is the average variance within individual groups.

TABLE 10–3
Calculation of S_M^2

	Group Mean	d	d^2
	1.0	−1.5	2.25
	2.0	− .5	.25
	3.0	+ .5	.25
	4.0	+1.5	2.25
Total	10.0	.0	5.00
Mean	2.50	.0	$1.25 = S_M^2$

The Breakdown of Sums of Squares. There is probably no real point in your doing it yourself, but, if you compute Σd_T^2 in Table 10–2 $(-2.5^2 + -2.5^2 + \cdots + 2.5^2 + 3.5^2)$, you will find that this total is 78. Computing Σd_{BG}^2 $(-1.5^2 + -1.5^2 + \cdots + 1.5^2 + 1.5^2)$ yields a total of 50. A similar calculation of Σd_{WG}^2 $(-1.0^2 + 1.0^2 + \cdots + 1.5^2 + 1.5^2)$ gives a total of 28. This shows in practical terms another fact mentioned earlier:

$$\Sigma d_T^2 = \Sigma d_{BG}^2 + \Sigma d_{WG}^2$$
$$78 = 50 + 28$$

Before I bring this section to a close, there is one final point to make. There is a very simple relationship between the variances calculated in the previous section and the sums of squares calculated in this one. In each case $\Sigma d^2 = NS^2$. Thus

$$NS_T^2 = \Sigma d_T^2 \qquad 40 \times 1.95 = 78$$
$$NS_{BG}^2 = \Sigma d_{BG}^2 \qquad 40 \times .70 = 28$$
$$NS_{WG}^2 = \Sigma d_{BG}^2 \qquad 40 \times 1.25 = 50$$

If you think back to the calculations, you will probably sense that this is because directly or indirectly all 40 scores enter into the determination of the value of each term.

The F Test. With Σd_{BG}^2 and Σd_{WG}^2 calculated, it is almost time to introduce the F ratio, the statistic computed for analysis of variance. Recall that this test involves a comparison of an estimate of population variance based on variation between-group means with another estimate based on average variation within individual groups. The first of these estimates is obtained from the formula

$$S_{BG}^2 = \frac{\Sigma d_{BG}^2}{K - 1}$$

where K is the number of groups in the study.

The second estimate is obtained from the formula

$$\hat{S}_{WG} = \frac{\Sigma d_{WG}^2}{N - K}$$

The formula for F is

$$F = \frac{\hat{S}_{BG}^2}{\hat{S}_{WG}^2}$$

Degrees of Freedom. Before I go further with the interpretation of F, I will need to take a moment to deal with one of the most fundamental concepts in statistics, the concept of *degrees of freedom*, abbreviated d.f. The concept is a difficult one, and all I will do here is to tell you that, for a set of data, the number of degrees of freedom is the number of terms that can be assigned freely. Suppose you know that the total of 10 numbers is 27. The first 9 of these 10 numbers can be anything, for example, 30, 35, 40, 45, 50, 55, 60, 65, and 70, yielding a sum for just 9 numbers of 450. This means that the last number *must be* -423 to arrive at a total of 27. This last number cannot be assigned freely. In this example there were 9 degrees of freedom, one less than the number of numbers.

The concept of degrees of freedom has come up once before in this book, although it seemed unnecessary to call attention to such unpleasantnesses then. This was in connection with the calculation of an unbiased estimate of σ (p. 137) by dividing Σd^2 by $N-1$ (degrees of freedom) rather than by N. As it turns out, degrees of freedom play a similar role in ANOVA because the F test comes down to a comparison of estimates of population variance.

To illustrate, let us return to the study of life stress and sickness. There were 40 participants in the hypothetical study. The total number of times they answered the sick call was 100. For the reasons given in the example of 10 numbers with a total of 27, the existence of this total of 100 also means that the number of degrees of freedom (d.f.) in the data is $N-1$ or $40-1=39$; an unbiased estimate of population variance would be $\Sigma d_T^2/39$.

One part of the analysis of variance involves the estimation of population variance on the basis of the means of the $K=4$ subgroups. Again the estimate costs a degree of freedom and the estimate is based upon $K-1=3$ d.f. *between groups*.

Of the original 39 d.f., 36 $(39-3)$ remain. Please note in passing that this is also $N-K=40-4$. These are the d.f. *within groups*. Each of the 4 subgroups of 10 individuals loses a degree of freedom because each has its own total. This leaves 9 d.f. in each and $9\times4=36$. A different estimate of population variance based upon the variance within the individual samples will have 36 d.f.

Now, at last, we are in a position to do the F test on the hypothetical life-stress and sickness data. The calculations require three steps.

$$\hat{S}^2_{BG}=\frac{\Sigma d^2_{BG}}{K-1}=\frac{50}{4-1}=16.67$$

$$\hat{S}^2_{WG}=\frac{\Sigma d^2_{WG}}{N-K}=\frac{28}{36}=.78$$

$$F=\frac{\hat{S}^2_{BG}}{\hat{S}^2_{WG}}=\frac{16.67}{.78}=21.37$$

For a good many purposes it is instructive to put all of this into a table. For the analyses just done, the data are in Table 10–4. In order each numbered column in the table: (1) identifies the separate components into which variance has been analyzed, (2) presents degrees of freedom for each component, (3) lists Σd^2, (4) gives an estimated population variance ($\hat{S}^2$) obtained by the formula $\Sigma d^2/\text{d.f.}$, (5) gives the value of the statistic, F, and (6) indicates the level of confidence with which the null hypothesis can be rejected.

TABLE 10–4
Typical ANOVA Table

(1) Source of Variation	(2) d.f.	(3)* Σd^2	(4) $\hat{S}^2$	(5) F	(6) P
Between-groups (stress-unit levels)	3	50	16.67	21.37	<.01
Within-groups	36	28	.78		
Total*	39	78			

*The row labeled "Total" and the column labeled "Σd^2" usually do not appear in these tables. I included them for the sake of making a couple of points a little clearer.

These last points require just a little further interpretation. The value of F is the ratio of some estimate of population variance based upon an effect of interest to an appropriate *error term,* which is also an estimate of population variance. This error term is not unlike the terms S_M and S_{diff} (pp. 141–154) which were involved in hypothesis testing using the Z test. The evaluation of F is also essentially like the evaluation of Z, in that tables are available which give the probability of obtaining an F of the size obtained if the null hypothesis is true. In other words, these tables refer the user to the *sampling distribution* of F for a determination of the probability of occurrence of a given value of F.

The big difference between the F distribution and the Z distribution is that the distribution of F is different for different numbers of d.f. Moreover, there are two values of d.f. to consider, that associated with the numerator and that associated with the denominator in the F ratio. A fragment of an F table appears as Table 10–5. To use it, you first locate the column for the number of degrees of freedom associated with the numerator in the F ratio. This is the number of degrees of freedom for the effect being evaluated. In this case, there were four groups with differing levels of stress whose means provided the between-groups estimate of population variance ($\hat{S}^2_{BG}$). Since there were four groups, this estimate has 3 d.f.

TABLE 10–5
Table of F Ratios Significant at $p=.05$ and $p=.01$

Degrees of Freedom in Denominator		Degrees of Freedom in Numerator				
		1	2	3	4	5
1	$p=.05$	161	200	216	225	230
	$p=.01$	4,052	4,999	5,625	5,625	5,764
5	$p=.05$	6.61	5.79	5.41	5.19	5.05
	$p=.01$	16.26	13.27	12.06	11.39	10.97
10	$p=.05$	4.96	4.10	3.71	3.48	3.33
	$p=.01$	.04	7.56	6.55	5.99	5.64
20	$p=.05$	4.35	3.49	3.10	2.87	2.71
	$p=.01$	8.10	5.85	4.94	4.43	4.10
30	$p=.05$	4.17	3.32	2.92	2.69	2.53
	$p=.01$	7.56	5.39	4.51	4.02	3.70
36	$p=.05$	4.11	3.26	2.86	2.63	2.48
	$p=.01$	7.39	5.25	4.38	3.89	3.58
50	$p=.05$	4.03	3.18	2.79	2.56	2.41
	$p=.01$	7.17	5.06	4.20	3.74	3.42
100	$p=.05$	3.94	3.09	2.70	2.46	2.30
	$p=.01$	6.90	9.82	3.98	3.51	3.20
∞	$p=.05$	3.84	2.99	2.60	2.37	2.21
	$p=.01$	6.64	4.60	3.78	3.32	3.02

Having located the appropriate column, the next step is to find the row in the table for the number of degrees of freedom of the denominator—the error term. In this case, this is $\hat{S}^2_{WG}$, which has 36 d.f. I have included the entries for 3 and 36 d.f. in the table: an F of 2.86 is significant at $p=.05$; an F of 4.38 is significant at the 1% level of confidence. The value of 21.37 (from the ANOVA table for the results) is much larger than either of these numbers and therefore very significant.

The t Test

The very simplest experiment of the type I have been describing involves just two groups with some number of participants in each group. Although the results of such bivalent experiments are usually evaluated by means of a statistic called the t test, they can also be handled by the methods of ANOVA. In such an analysis, d.f. for experimental conditions ($\hat{S}^2_{BG}$) is always 1, because there are just two groups. The d.f. for the error term is the total number of subjects in the two groups minus 2, that is, $N-K$.

I shall describe the calculations required to do a t test in the computational appendix (p. 238). There is no need to do anything special here, however, because of a certain relationship of F to t.

$$F = t^2 \quad \text{and} \quad t = \sqrt{F}$$

Moreover, when the d.f. are infinite, t is equal to Z. Table 10–5 allows you to make a partial check on this point. If you take the square roots of the values entered for 1 and ∞ d.f., you obtain 1.96 and 2.58, exactly the values of Z required to reject the null hypothesis at the 5% and 1% levels of confidence.

FACTORIAL DESIGNS

The methods presented so far in this chapter are often called *one-way analysis of variance* because they assess the effects of just one independent variable, life stress in the example I have been dealing with. When the methods are applied to more complex designs, they become "many-way" analyses because of the additional independent variables. In this book I have limited myself to *two-way analysis of variance*, applied to factorial designs of the type presented in Chapter 4.

Experimental Design

For starters, it may make sense for you to review the topic of factorial designs in Chapter 4 (pp. 70–75). If you remember those materials pretty well, these facts may serve as an adequate reminder: (1) in a factorial experiment there are two or more independent variables, crossed so that each value of every variable is combined with each value of every other variable; and (2) these experiments yield (a) terms that reflect the *main effects* of each independent variable and (b) other terms that show the *interaction* between or among the independent variables. My task in this section will be to describe the analysis of such experiments. As a way of presenting such analyses, I will deal with the same set of data four different times, imagining in each case that four subgroups of these data have different meanings.

An Example

Suppose that you are the publisher and distributor of *Cosmic Encyclopedia*, a publication that finds its way chiefly into the libraries of secondary schools and colleges. Suppose further that you are about to bring out a new edition of *CE* and are beginning to worry about the problems of promotion. In one region of the country your sales staff consists of four teams of 10 people each. In the last year of the current edition, their individual performances were as shown in Table 10–6 in terms of the number of libraries that bought the encyclopedia. The numbers in the table are the numbers of sales by each member of each 10-person team. The numbers are deliberately idealized for the purpose of making my later points.

TABLE 10–6
Sales of Encyclopedias by Four Teams

	Team				
	A	B	C	D	Total Group
	20	40	40	20	
	20	40	40	20	
	20	40	40	20	
	30	50	50	30	
	30	50	50	30	
	30	50	50	30	
	30	50	50	30	
	40	60	60	40	
	40	60	60	40	
	40	60	60	40	
Mean	30	50	50	30	40
S^2	60	60	60	60	160.00
Σd^2	600	600	600	600	6,400.00

From the means (30, 50, 50, 30) it is obvious that there are differences in the sales of the four groups. A question to ask immediately is whether these differences are chance variations or whether they are real. The answer can be had by ANOVA. I will use this analysis to review the calculations involved in one-way analysis—partly just for the sake of review but, more important, because the analysis of factorial experiments begins with the first steps of a one-way analysis.

One-Way ANOVA Revisited. The terms we need to do a one-way analysis of these data are two estimates of population variance, one ($\hat{S}_{BG}$) based upon the variation of group means, the other ($\hat{S}_{WG}$) based on the variation within groups. The first of the estimates is given by the formula

$$\hat{S}^2_{BG} = \frac{\Sigma d^2_{BG}}{K-1}$$

where K is the number of groups. Since $d_{BG}^2 = NS_M^2$, one way to obtain this value is to compute the variance of the means and multiply by $N=40$. Thus

$$S_M^2 = \frac{(30-40)^2+(50-40)^2+(50-40)^2+(30-40)^2}{4}$$

$$= \frac{400}{4} = 100$$

$$\Sigma d_{BG}^2 = 40 \times 100 = 4{,}000$$

Within-groups variance S_{WG}^2 is the average of the variances of the individual groups. Since they all have the same value, 60, S_{WG}^2 will be 60 and

$$\Sigma d_{WG}^2 = 40 \times 60 = 2{,}400$$

Now we are in a position to make the necessary estimates of population variance.

$$\hat{S}_{BG}^2 = \frac{4{,}000}{3} = 1{,}333.33$$

$$\hat{S}_{WG}^2 = \frac{2{,}400}{36} = 66.67$$

and

$$F = \frac{1{,}333.33}{66.67} = 20$$

Consulting Table 10-5, it turns out that this value of F is significant well beyond the 1% level of confidence.

Two-Way Analysis of Variance

The management of *Cosmic* now has the problem of understanding these results and perhaps profiting from them in the new sales campaign. Suppose that two of the sales teams had concentrated on secondary schools and the other two had concentrated on colleges. Suppose further that two of the sales teams were all women and the other two were all men. It could be that one or both of these factors is responsible for the differences. I am deliberately not saying which performances are related to which variable because I want to reanalyze the data with different possibilities in mind.

To begin with, let us consider what some of these possibilities are: (1) female teams are more effective than male teams, or vice versa: there is a main effect of sex: (2) secondary schools buy more encyclopedias than colleges, or vice versa: there is a main effect of type of school; (3) men (or women) are more effective in colleges (or secondary schools), or vice versa: there is an interaction between sex and type of school; or (4) a combination of the above: for example, women are more effective than men in general (main effect) but the difference is less in secondary schools than in colleges (interaction).

One thing that may help your understanding is the following observation. The means of these four groups are what they are: 30, 50, 50, and 30. The variance among them must contain the differences due to sex, type of school, and/or an interaction. In short, the between-groups variance is an "all-causes" variance. To put it another way, the between-groups variance can be further analyzed into components of variance related to these influences. Once this is done, estimates of population variance are possible, and these estimates may be tested against the within-groups estimate, just as in the original analysis. Now for some examples.

Each of the examples will be a 2×2 factorial arrangement of the means for the different sales teams. In each case the arrangement will be like that in Table 10–7, where I have also entered means for the four groups, identified by the letter that went with the group in the original table.

TABLE 10–7

	Sex		
Type of School	Male	Female	Mean
College	A 30	B 50	40
Secondary	D 30	C 50	40
Mean	30	50	

This set of results says that groups A and D were males, groups B and C females, groups A and B worked in colleges, and groups C and D worked in secondary schools. The data tell us that females are more effective salesmen than males are salespersons (means of 50 versus 30), but that type of school has no effect (means of 40 and 40). Thus the variance associated with type of school is zero. But now suppose that we compute the variance for sex. The overall mean for men and women is 40. So:

$$\frac{\Sigma d^2}{N} = \frac{(30-40)^2+(50-40)^2}{2}$$

$$= \frac{10^? + 10^? =}{2} = \frac{100+100}{2} = 100$$

If you check back on my earlier computations, you will find that S_M^2 in the one-way analysis was also 100. In this case, the between-groups variance is a variance associated with sex. Think this through, recalling that between-groups variance is an "all-causes" variance. Since one variable is contributing zero to this variance, the other variable accounts for all of it.

Suppose that we fill in the table as shown in Table 10–8. This arrangement of the means would tell us that the entire between-groups variance is associated with type of school. The calculation of this variance would be exactly the same as for the previous case and the variance for type of school would be 100.

TABLE 10–8

	Sex		
Type of School	Male	Female	Mean
College	A 30	D 30	30
Secondary	B 50	C 50	50
Mean	40	40	

To do an F test for either of these analyses we need to get back to $\Sigma d_{BG}^2 = NS_M^2 = 40 \times 100 = 4{,}000$. This term has only one d.f. (2 groups minus 1), so $\hat{S}_{BG}^2$ is also 4,000, and evaluated by the S_{WG}^2 calculated earlier,

$$F = \frac{\hat{S}_{BG}^2}{\hat{S}_{WG}^2} = \frac{4{,}000}{66.67} = 60$$

Finally, suppose that the means of the four groups arrange themselves as in Table 10–9. In this case there is no main effect of sex and no main effect of type of school, but there is a substantial interaction: males are more effective in secondary schools; females are more effective in colleges.

TABLE 10–9

Type of School	Sex		Mean
	Male	Female	
College	A 30	D 50	40
Secondary	C 50	D 30	40
Mean	40	40	

The easiest way to arrive at a value for the variance associated with this interaction is by subtracting sums of squares. This will also give me an opportunity to do another brief review of materials covered earlier. Recall that

$$\Sigma d_T^2 = \Sigma d_{BG}^2 + \Sigma d_{WG}^2$$

The between-groups sum of squares is an "all-causes" sum of squares, however, made up of three components, in this case

$$\Sigma d_{BG}^2 = \Sigma d^2_{\substack{\text{type of}\\ \text{school}}} + \Sigma d_{\text{sex}}^2 + \Sigma d_{\text{interaction}}^2$$

This means that the sum of squares for interaction can be obtained this way:

$$\Sigma d_{\text{interaction}}^2 = \Sigma d_{BG}^2 - \Sigma d^2_{\substack{\text{type of}\\ \text{school}}} - \Sigma d_{\text{sex}}^2$$

If you recall from previous analyses that Σd_{BG}^2 was 4,000 and that in this case the two sums of squares for main effects are zero, this means that

$$\Sigma d_{\text{interaction}}^2 = 4{,}000 - 0 - 0$$
$$= 4{,}000$$

Thus a variance estimate base on the interaction is the sum of squares divided by its single d.f.:

$$\hat{S}_{\text{interaction}}^2 = \frac{4{,}000}{1} = 4{,}000$$

and

$$F = \frac{4{,}000}{66.67} = 60$$

TABLE 10–10

Source of Variation	d.f.	Estimated σ^2	F	p
Sex	1	0	0	
Type of school	1	0	0	
Sex × type of school	1	4,000	60	$<.01$
Within groups	36	66.67		

For this case I can make one additional point by constructing the ANOVA table, Table 10–10. The additional point is that the analysis of between-groups variance into three components also divided its degrees of freedom into three components. To summarize this bit of information, the degrees of freedom in a total set of data are $N-1$. In an analysis involving only between- and within-groups variances for K groups, the degrees of freedom are $K-1$ for the between-groups estimate and $N-K$ within groups. If the analysis proceeds further, as is possible in a factorial experiment, by breaking down the between-groups variance into its components there are $A-1$ d.f. for a first variable, $B-1$ d.f. for a second, and $(A-1)\times(B-1)$ for the interaction. Moreover, $A-1+B-1+[(A-1)\times(B-1)]=K-1$.

SUMMARY–GLOSSARY

In complex experiments involving more than two sets of data a problem of analysis arises. Methods that compare just two individual means at a time are inappropriate because such comparisons would not be independent. The methods of analysis of variance deal with this problem by testing the null hypothesis that the variance of the means is not too great for the set of means to have come from a population whose variance can be estimated from the average variance within groups. Analysis of variance begins by breaking the total variance in a set of data down into two or more components that lead to different estimates of population variance. Rephrased, the null hypothesis tested by analysis of variance is that these estimates are no different. The key concepts covered in this chapter are the following.

ANOVA. An acronym for *analysis of variance*.

One-way analysis of variance. An analysis of variance performed on the results of a multivalent experiment which involves the manipulation of a single independent variable.

Two-way analysis of variance. The analysis of variance appropriate to a factorially designed experiment with two or more independent variables.

t-test. The statistical test appropriate to a two-group experiment. t is related to the F ratio in that $t^2 = F$.

F test. Also called the *F ratio,* this is the statistical test used in analysis of variance. A main effect or interaction (see below) is divided by an appropriate error variance to conduct this test. The result is referred to a table of probabilities of obtaining an F as large as that obtained. The null hypothesis is rejected if this probability is sufficiently low.

Main effect. The effect of variations in one variable in a factorial experiment when these effects are collapsed across all values of other variables in the experiment.

Interaction. The effect of manipulating one variable changes with changes in the values of others.

Degrees of freedom (d.f.). The number of terms in a calculation that can be assigned freely.

Between-groups variance. An estimate of population variance based on the variation among group means.

Within-group variance. An estimate of population variance based on the average variation among the measures within samples. In the F tests discussed in this chapter, this is error variance.

Computational Appendix

The Puritan Ethic dies hard in our culture.

There was a day when Farmer Hiram Sample was sitting on the rail fence that separated his farm from that of Lucius Random, who owned the neighboring property. He sat there wearing a look of disbelief and bewilderment at what was going on next door: Farmer Random had his prize bull, the sire of many stock show champions, harnessed to a plow and with the aid of an enormous whip was turning sod at a furious rate. Finally, curiosity won out and Hiram called across the field. "Luke, what in tarnation are ya doin' there?"

"Plowin'," came the laconic reply.

"Yeah, I can see that," Hiram said, "but beatin' the b'jeezus outa the critter with the big whip. He'll be ruint for makin' calves forever. How come you're doin' it to him?"

Lucius answered, "Just teachin' the son of a bitch that there's more to life than makin' love!"

I suppose that an attitude of the type caricatured in this little scenario plays more of a part than it ought to in the approaches of teachers and students toward statistics. No doubt teachers more than students, but a common philosophy appears to be that suffering somehow ennobles a person and is a value to be prized for its own sake.

Probably this philosophy is right, but right and necessary are two different things. My whole effort in this book has been to try to show you that gaining an understanding of statistical thinking can be a painless undertaking. If I have been even halfway successful, I suspect that this is because I have kept formulas and computational mechanics to a minimum.

I turn now to a very brief summary of the procedures that are actually used in statistical analysis and some of the derivations that lie behind them. My bet is that you will find them a very simple matter now that you have the sense of the concepts served by these calculations.

DESCRIPTIVE STATISTICS

The first part of what I want to cover is almost at the level of a review of vocabulary. For that reason I shall be very brief.

The Mean

Let X stand for each individual score, N for the number of such scores, and Σ for the process of summation. Then the formula for the mean, M, is

$$M = \frac{\Sigma X}{N} \tag{1}$$

If scores are grouped as they were in the example in Chapter 6 involving the mean number of children in 100 families, the formula for the mean is slightly different:

$$M = \frac{\Sigma fX}{N} \tag{2}$$

where f stands for the frequency in each of the several groups of scores. Table A–1 repeats the example.

TABLE A–1

Number of Children, X	Number of Families, f		fX
0	6	0+0+0+0+0+0	0
1	9	↓	9
2	16		32
3	14		42
4	13		52
5	11		55
6	9		54
7	7		49
8	5	8+8+8+8+8	40
9	4	9+9+9+9	36
10	3	10+10+10	30
11	2	11+11	22
12	1		12
Total	$100 = N$		$433 = \Sigma fX$

$$M = \frac{\Sigma fX}{N} = \frac{433}{100} = 4.33$$

The revealing part of the table is in the middle, which shows that the expression fX is the equivalent of adding the scores in that category, so that this formula is the same as formula (1).

Proof That $\Sigma d = 0$

In the text I went through a long rigamarole to show you that, sign of d respected, the sum of the deviations from the mean is zero. The algebraic demonstration is much quicker.

$$d = X - M$$

$$\Sigma d = \Sigma X - \Sigma M \qquad \text{since } M \text{ is constant}$$

$$= \Sigma X - NM \qquad \text{since } M = \Sigma X / N$$

$$= \Sigma X - N\frac{\Sigma X}{N} \qquad \text{canceling } N$$

$$= \Sigma X - \Sigma X = 0$$

Percentile Rank

Let the expression Σf stand for *cumulative frequency*, the process of adding the frequencies of the scores successively from the bottom of a distribution. Then

$$\text{percentile rank} = 100 \times \frac{\Sigma f}{N} \qquad (3)$$

Table A–2 illustrates everything involved for $N = 50$ scores ranging from 1 through 10. The percentile ranks of each of the 10 scores appear in the right-hand column. These calculations (and the formula) implicitly define the percentile rank of a score as *the percentage of cases which that score equals or exceeds.*

TABLE A–2

Score	f	Σf	$\Sigma f/N$	$(\Sigma f/N) \times 100$
10	4	50	1.00	100
9	6	46	.92	92
8	5	40	.80	80
7	8	35	.70	70
6	9	27	.54	54
5	6	18	.36	36
4	5	12	.24	24
3	3	7	.14	14
2	2	4	.08	8
1	2	2	.04	4

Cumulative Frequency Distributions

If you were to plot the numbers in the column headed Σf against scores from 1 through 10, the result would be a cumulative frequency distribution which sometimes figures in the graphical solution of statistical problems. Figure A–1 is the cumulative frequency distribution of the data in the preceding section. Two examples will show what you can do with such a plot. I shall describe one of these applications now, the other in the next section. The first example: To determine the percentile rank of a score of 7, just read up to the cumulative frequency distribution from 7 and then over to the right-hand ordinate, labeled "percentile rank." You discover that the percentile rank of a score of 7 determined in this way is 70. Checking back to the table, you see that this answer is correct.

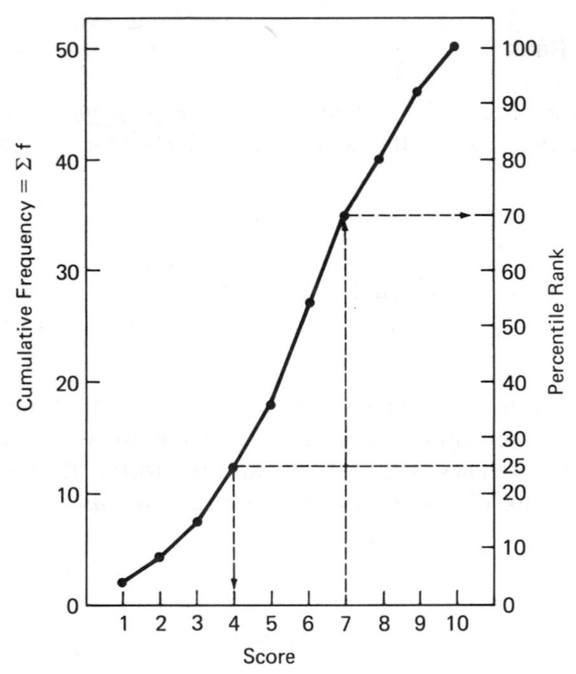

Figure A–1 Cumulative frequency distribution. This figure presents the data appearing in the text in tabular form.

Score at a Given Percentile Rank

Turning the procedure around and asking what score corresponds to a given percentile rank raises a small problem, because the exact score you want may not appear in the cumulative frequency distribution. If you wanted to determine the score at the 25th percentile in our example, you would find that no score in the table is exactly at that percentile rank. A score of 4 is at the 24th percentile and a score of 5 is at the 36th percentile, so the 25th percentile must be a little more than 4. Apparently the answer will be 4 point something or other.

One way to determine the "something or other" is with a cumulative frequency distribution, and I have shown the procedure in Figure A-1. If you read from the right-hand ordinate over to the curve and then down to the abscissa, you will see that the value there appears to be about 4.1.

Another approach to the answer would be to reason that, since the percentile rank of a score of 4 is 24 and the percentile rank of a score of 5 is 36, the score you want is 1/12 of the distance from 4 to 5, or 4.08 = 4.1.

There are formulas which give you a more precise answer to the question of the score corresponding to a given percentile rank. They involve such matters as the real limits of the class intervals into which the scores are grouped, all of which I find a terrible bore. In my opinion the two procedures I have described are precise enough. In actual practice percentiles most often figure in statements of rough approximations.

The Median

Since the median is the middle score, the 50th percentile, the procedures to determine it are the same as for calculating the score corresponding to other percentile ranks. You may want to determine it for the scores we have been discussing by the graphic method and the method of interpolation I have just described. I get about 5.8 by both methods.

The Standard Deviation (S)

Let d stand for $X - M$, the deviation of a score from the mean. There will be one such d for every X. In these terms,

$$S = \sqrt{\frac{\Sigma d^2}{N}} \tag{4}$$

Variance (S^2)

Since variance is the square of the standard deviation,

$$S^2 = \frac{\Sigma d^2}{N} \tag{5}$$

Calculating Formulas for S and S^2

The calculation of S and S^2 by these formulas is pretty painful. They are very useful for communicating the meaning of the concepts but are almost never used in practical situations. The derivation of calculating formulas is easy.

1. $S^2 = \frac{\Sigma d^2}{N}$ and $d = X - M$; thus $d^2 = (X - M)^2$

2. $d^2 = (X - M)^2 = X^2 - 2XM + M^2$

3. $\frac{\Sigma d^2}{N} = \frac{\Sigma X^2}{N} - \frac{2\Sigma XM}{N} + M^2$

3a. $\frac{\Sigma X}{N} = M$; thus $\frac{2\ \Sigma X\ M}{N} = 2M \cdot M = 2M^2$

4. $\frac{\Sigma d^2}{N} = \frac{\Sigma X^2}{N} - 2M^2 + M^2$

5. $\frac{\Sigma d^2}{N} = \frac{\Sigma X^2}{N} - M^2$; thus

6. $$S^2 = \frac{\Sigma X^2}{N} - M^2 \qquad (6)$$

7. $$S = \sqrt{\frac{\Sigma X^2}{N} - M^2} \qquad (7)$$

In Chapter 6 I reported that the standard deviation of the weights of 17 Miss Americas was 5.60. I shall use this example to illustrate the calculations and to make a point that sometimes has a great deal of practical usefulness.

This point is that adding or subtracting a constant from every score in a distribution does not change the standard deviation or variance. It does increase or reduce the mean by the constant added or subtracted from each score.

Dividing each score by a constant decreases the mean and the standard deviation by that constant. Multiplying by a constant has the opposite effect.

Either of these methods of transforming scores is sometimes of considerable usefulness for the purpose of simplifying calculations.

Consider first only the first two columns in Table A–3, which present each value of X (weight), its square, and at the bottom totals and means for these measures. The latter values actually enter the equations.

TABLE A–3

	Weight, X	X^2	$X-100$	$(X-100)^2$
	114	12,996	14	196
	120	14,400	20	400
	116	13,456	16	256
	118	13,924	18	324
	115	13,225	15	225
	124	15,376	24	576
	124	15,376	24	576
	115	13,225	15	225
	116	13,456	16	256
	135	18,225	35	1,225
	125	15,625	25	625
	110	12,100	10	100
	121	14,641	21	441
	118	13,924	18	324
	120	14,400	20	400
	125	15,625	25	625
	119	14,161	19	361
Total	2,035	244,135	335	7,135
Mean	119.706	14,360.882	19.706	419.706

$$S=\sqrt{\frac{\Sigma X^2}{N}-M^2}$$

$$=\sqrt{\frac{244,135}{17}-(119.706)^2}$$

$$=\sqrt{14,360.882-14,329.526}$$

$$=\sqrt{31.356}=5.600$$

Now let us carry out the same calculations for the numbers in the third and fourth columns of the table, which involve the subtraction of 100 pounds from each weight in column 1.

$$S=\sqrt{\frac{\Sigma X^2}{N}-M^2}$$

$$=\sqrt{\frac{7,135}{17}-(19.706)^2}$$

$$=\sqrt{419.706-388.326}$$

$$=\sqrt{31.380}=5.602$$

The obvious thing, of course, is that the answers are about the same in the two calculations. In fact, adding or subtracting a constant from each score in a set of scores reduces the mean by that constant but does not alter the variance or the standard deviation at all. The fact that the two calculations gave slightly different answers ($S^2=31.356$ versus 31.380; $S=5.600$ versus 5.602) results from nothing more than the different effects of rounding on small and large numbers.

A couple of alternative calculating formulas are also available. It will be worthwhile deriving them because they will come up again, once almost immediately and again later on in connection with analysis of variance. The derivation involves getting an expression for Σd^2 in the second step of the process.

1. $\frac{\Sigma d^2}{N}=\frac{\Sigma X^2}{N}-M^2$; multiplication by N yields
2. $\frac{\not{N}\Sigma d^2}{\not{N}}=\frac{\not{N}\Sigma X^2}{\not{N}}-NM^2$; thus
3. $\Sigma d^2=\Sigma X^2-NM^2$ and
4. $S^2=\frac{\Sigma X^2-NM^2}{N}$ (8)
5. $S=\sqrt{\frac{\Sigma X^2-NM^2}{N}}$ (9)

A slightly different and often more useful formula substitutes $(\Sigma X)^2/N$ for NM^2.

1. $NM^2=N\left(\frac{\Sigma X}{N}\cdot\frac{\Sigma X}{N}\right)$ by definition
2. $NM^2=\frac{\not{N}(\Sigma X)^2}{\not{N}^2}=\frac{(\Sigma X)^2}{N}$, and since (10)
3. $S^2=\frac{\Sigma X^2-NM^2}{N}$ and $NM^2=\frac{(\Sigma X)^2}{N}$
4. $S^2=\frac{\Sigma X^2-(\Sigma X)^2/N}{N}$ (11)
5. $S=\sqrt{\frac{\Sigma X^2-(\Sigma X)^2/N}{N}}$ (12)

Estimates of σ and σ^2

Recall that $\hat{S}$ and $\hat{S}^2$ are the symbols for estimates of σ and σ^2, respectively. In previous discussions I have used these deviation formulas to arrive at these estimates:

$$\hat{S}^2 = \frac{\Sigma d^2}{N-1} \tag{13}$$

$$\hat{S} = \sqrt{\frac{\Sigma d^2}{N-1}} \tag{14}$$

Applying the derivations in the previous sections, it is clear that the calculating formulas will be

$$\hat{S}^2 = \frac{\Sigma X^2 - NM^2}{N-1} \quad \text{or} \tag{15}$$

$$= \frac{\Sigma X^2 - (\Sigma X)^2/N}{N-1} \tag{16}$$

$$\hat{S} = \sqrt{\frac{\Sigma X^2 - NM^2}{N-1}} \quad \text{or} \tag{17}$$

$$= \sqrt{\frac{\Sigma X^2 - (\Sigma X)^2/N}{N-1}} \tag{18}$$

It may be worth a moment to think back to the S of 5.60 calculated for the weights of all the Miss Americas to make the point that if one's interest is in the variability of weights of previous Miss Americas, this value is also σ, because the 17 women *are* the population. For such purposes the $N-1$ formula would be inappropriate. If you wanted to use the data as an estimate of the variability of all Miss Americas, including those to be chosen in the future, and were willing to assume that nothing will happen to judges' tastes to affect this variability (a risky assumption), formula (16) would apply. Going back to the table for the essential information,

$$\hat{S}^2 = \frac{\Sigma X^2 - (\Sigma X)^2/N}{N-1}$$

$$= \frac{244{,}135 - (2{,}035)^2/17}{16}$$

$$= \frac{244{,}135 - 4{,}141{,}225/17}{16}$$

$$= \frac{533.53}{16} = 33.35$$

$$\hat{S} = \sqrt{33.35} = 5.77$$

With these numbers handy, I can illustrate the application of still another set of formulas for $\hat{S}^2$ and $\hat{S}$.

$$\hat{S}^2 = \frac{N}{N-1} S^2 \tag{19}$$

$$\hat{S} = \sqrt{\frac{N}{N-1}}\; S \tag{20}$$

Recall that $S = 5.6021$ and $\hat{S}^2 = 31.3841$. The four decimal places are to make this work exactly. Thus

$$\hat{S}^2 = \frac{17}{16} \times 31.3841$$

$$= 1.0625 \times 31.3841 = 33.3456$$

$$\hat{S} = \sqrt{33.3456} = 5.77$$

Alternatively,

$$S = \sqrt{\frac{17}{16}} \times 5.6021$$

$$= 1.0625 \times 5.6021$$

$$= 1.0308 \times 5.6021$$

$$= 5.77$$

COUNTING STATISTICS

It is useful to divide statistical procedures up into two broad categories, those based on measurement and those based on counting. Of the terms reviewed so far in this appendix, M, S, $\hat{S}$, S^2, and $\hat{S}^2$ involve measurement. The median and other percentiles involve counting. The statistics associated with probability theory are also counting statistics. In this section I shall add a bit to the materials on probability and also introduce you to a new test, χ^2 (chi-square), which is the most important of the tests in this category.

Probability

Let p stand for the probability of an event of interest, q for $1-p$, and N for the total number of events under consideration. In these terms probabilities for heads and tails are $p = .5$ and $q = .5$. For a toss of unbiased coins the expected number of heads in 100 throws is given by a formula for the mean

$$M = Np \qquad (21)$$
$$100 \times .5 = 50$$

When N is large, the curve of chance outcomes approximates the normal curve and it is sensible to compute a standard deviation for the number of heads by the formula

$$\begin{aligned} S &= \sqrt{Npq} \\ &= \sqrt{100 \times .5 \times .5} = \sqrt{100 \times .25} = \sqrt{25} \qquad (22) \\ &= 5.0 \end{aligned}$$

Since the formula for S is available, it is possible to compute Z scores and to use them to evaluate the probability of an outcome against expectations. I used such a procedure to evaluate the probability of Mr. Z's visits in Chapter 1. Recall that there were 30 visits altogether, but that they split 21 to 9 when the expected division was 15 to 15.

$$\begin{aligned} M &= Np = 30 \times .5 = 15 \\ S &= \sqrt{Npq} = \sqrt{30 \times .5 \times .5} = \sqrt{7.50} = 2.74 \\ Z &= \frac{X - M}{S} \qquad (23) \\ &= \frac{21 - 15}{2.75} = \frac{6}{2.75} = 2.19 \end{aligned}$$

which has a probability of less than .05. It may be worth noting that the calculations and the answer would have been the same, except for sign, if the computations had been done for the 9 visits instead of the 21.

The probability that the Boston Clerk of Court would choose only 90 women in a set of 300 potential jurors by chance can be approached in the same way. Give him the benefit of the doubt to start with and assume that there are no more women than men in the Boston area, so that the probability of selecting a woman is .5. Then the expected number of women is

$$M = Np = 300 \times .5 = 150$$

The standard deviation in this case is

$$S = \sqrt{Npq} = \sqrt{300 \times .25} = \sqrt{75} = 8.66$$

and the Z test proceeds as follows:

$$Z = \frac{90-150}{8.66} = \frac{-60}{8.66} = 6.93$$

The odds against the occurrence of a Z this large by chance are as mentioned on page 53, only 1 in many million.

The Sign Test

The logic of the test just described is very close to that of a test that can be used in any case where data can be classified as positive (+) and negative (−) outcomes. Suppose that some political candidate wants to evaluate the effectiveness of a speech he makes in favor of liberalizing abortion laws. Prior to a test, a survey research agency locates 100 people who say that they are neutral on the subject and are willing to listen to a TV discussion of the issue. Then, as a test, the candidate makes the speech and the agency contacts the 100 viewers again, asking whether their reactions were affected one way or another. Ten of the respondents say that they missed the speech or that it had no effect; 55 say that they were influenced positively (+) and 35 say that they were influenced negatively (−). It is because of this possibility of assigning signs to outcomes that the test is sometimes called the *sign test*. The first thing we have to do to analyze these data is to decide what to do about the 10 unaffected viewers. By convention they are treated as "no observation" and dropped from the calculations. This reduces N to 90. If the direction of influence were random in these 90 people, the probability of a positive reaction would be .5 and the expected number of people reacting positively would be

$$M = Np = 90 \times .5 = 45$$

The standard deviation would be

$$S = \sqrt{Npq} = \sqrt{90 \times .5 \times .5} = \sqrt{22.50} = 4.74$$

Then, performing the same test as before,

$$Z = \frac{X-M}{S} = \frac{55-45}{4.75} = 2.11$$

A Z this large occurs by chance less than 5% of the time. On this basis the politician may reject the null hypothesis in favor of the hypothesis that his speech is effective and continue to use it. But he should be careful.

Perhaps the result occurred because the people surveyed were flattered at having been singled out for attention and responded as they thought the candidate wanted them to. In this case, the results reflect the demand characteristics of the study.

The χ^2 Test

One way to put what goes on in a test such as the sign test is that it compares the obtained frequency of some event with an expected frequency. The χ^2 test does the same thing except that it does it for several categories. Take this example: One bit of data collected in the ESP experiment described on page 97 was the number of heads in 5 guesses reported by the 100 students who participated in the demonstration. The expected frequencies of these events calculated on the basis of a binomial expansion appear in the third column of Table A–4. The obtained frequencies are in the second column. The formula for χ^2 is

$$\chi^2 = \sum_{1}^{k} \frac{(f_o - f_e)^2}{f_e} \qquad (24)$$

where χ^2 = chi-square
f_o = observed frequency of each outcome
f_e = expected frequency of each outcome
$\sum_{1}^{k}$ = signifies that the addition is over k categories: in this case, 6 (0, 1, 2, 3, 4, and 5 heads)

TABLE A–4

Number of Heads	Obtained Number of Students, f_o	Expected Number of Students, f_e	$(f_o - f_e)^2$	$\frac{(f_o - f_e)^2}{f_e}$
0	0	3	9	3.
1	13	16	9	.56
2	33	31	4	.13
3	44	31	169	5.45
4	9	16	49	3.06
5	1	3	4	1.33
Total	100	100		$\chi^2 = 13.53$

The last two columns in the table perform the indicated calculations and the total of the last column is χ^2. The significance of χ^2 is evaluated in terms of its degrees of freedom, which are $k-1$, in this case 5. Reference to appropriate tables reveals that this value of χ^2 is significant at between the 1% and 2% levels of confidence. This means that students' patterns of guessing heads and tails is not what one would expect on the basis of randomness.

The analysis just completed is slightly inappropriate because f_e in each category should be at least 5 to apply this method. I went ahead and did it, however, because I wanted to illustrate the procedure and did not want to confuse things. Moreover, a more appropriate analysis of the type I will use in my next example slightly increased the level of significance to beyond the 1% level of confidence, as mentioned on page 97.

The second example I will analyze involves the correctness of ESP guesses. The data appear in Table A–5, together with the calculation of χ^2. A standard table of χ^2 reveals that this result is not significant by accepted standards. An efficient way to express it is $.20>p>.10$, which says that, with 6 d.f.[1] a χ^2 of 10.20 is significant between the 10% and 20% levels of confidence.

TABLE A-5

Number Correct		f_o	f_e*	f_e-f_o	$(f_e-f_o)^2/f_e$
0		0			
1	(1)	0	5	4	3.20
2		1			
3		5	12	7	4.08
4		26	20	6	1.80
5		29	25	4	.64
6		22	20	2	.20
7		11	12	1	.08
8		3			
9	(6)	3	5	1	.20
10		0			
		100	99*		$\chi^2=10.20$

*Combining frequencies results in the total less than 100.

[1]Note in this case that I have combined categories 0, 1, 2 and 8, 9, 10 in order to obtain a f_e of at least 5 in every category. This reduces d.f. from 10 to 6 (7 categories -1). The total f_e is 99 rather than 100, as a result of rounding. Incidentally, don't read anything magical into the fact that a χ^2 of 10.20 is significant between $p=.10$ and $p=.20$. It is simply an accident.

HYPOTHESIS TESTING WITH CONTINUOUS VARIABLES

This section deals with the statistics for testing hypotheses about data based on continuous variables. Before I get into it, there is an important point to make. So far in this book I have made extensive use of Z scores. This was because I could illustrate the logic of hypothesis testing with the aid of a familiar statistic. One complicating point cannot be postponed any longer, however. It is, as I mentioned in Chapter 6 (p. 128), that Z scores can be interpreted accurately only if they refer to a normally distributed population of some sort.

One place where this condition fails to be met is with sampling distributions of means and differences whose parameters are estimated on the basis of small samples. The problem is that even using the unbiased estimate of σ, the distribution of Z scores has too many cases in the tails of the distribution. The situation is illustrated in Figures A–2 and A–3, where you can see two things: (1) as just mentioned, the calculation $(X-M)/S$ does produce more highly discrepant values than would be expected on the basis of a normal curve; and (2) the extent of this effect decreases with increasing numbers of cases or degrees of freedom.

Rather than inventing a substitute calculation for $(X-M)/S$, the solution to this problem is to refer the obtained value of $(X-M)/S$ to the sampling distribution that actually occurs with the number of d.f. involved. When this is done, the resulting statistical test is called the t test. From here on I shall use this test rather than Z, although you should understand

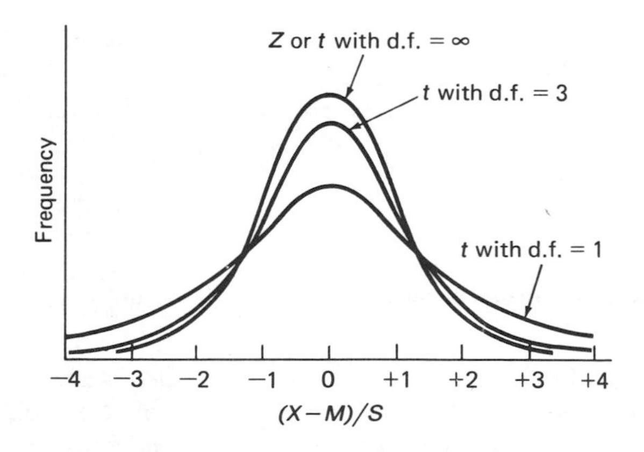

Figure A–2 Approximate sampling distributions of Z and t. The main things to get out of this figure are the following: (1) By comparison with the normal curve (Z), the t distribution contains fewer low values and more high values. This means that any value of t that would be significant at a particular level if it were Z is less significant. Figure A–3 amplifies on this point. (2) The extent of this effect decreases as d.f. increase.

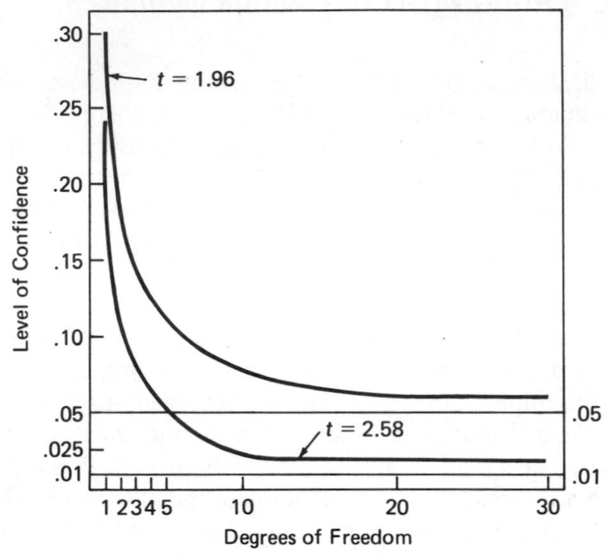

Figure A–3 Significance of t as a function of d.f. In the normal curve, Z values of 1.96 and 2.58 are significant at the .05 and .01 levels of confidence. In the case of t, the significance of these values depends upon d.f. One thing to notice is that by the time d.f. reach 20 or so, the downward slope of these functions is no longer detectable. They are asymptotic to .05 and .01 and d.f. $= \infty$. Beyond d.f. $= 20$, however, additional observations add very little to the power of the test, and what they do add comes very, very slowly. For this reason, experimental studies frequently have ns of 10–30 in each group. Since in the case of t, d.f. $= n_1 + n_n - 2$, not much is to be gained by adding more cases.

that the calculations are exactly the same. The only difference is that results are evaluated by referring to a table of t values rather than to a table of Z scores. Moreover, as I mentioned in Chapter 10, for large numbers of d.f., Z and t are very similar numbers; at infinity, they are identical.

Hypotheses About the Mean

Suppose you have reason to believe that the average IQ in your community is higher than the population average of 100. You collect 100 of your neighbors' IQs and obtain a mean of 105 and a standard deviation of 17 on this sample. To perform the test, you proceed as follows:

1. Calculate an unbiased estimate of the population standard deviation either by going back to your original data and using formula (14), (17), or

(18), or compute it by formula (20), as follows:

$$\hat{S}=\sqrt{\frac{N}{N-1}}\ S \tag{20}$$

$$=\sqrt{\frac{100}{99}}\times 17=1.005\times 17=17.2$$

2. Compute the standard error of the mean by the formula

$$S_m=\frac{\hat{S}}{\sqrt{N}} \tag{25}$$

$$=\frac{17.2}{\sqrt{100}}=\frac{17.2}{10}=1.72$$

This number is the standard deviation of the sampling distribution of the mean. Now you ask: if the population mean (μ) for my community is 100, the same as the general population, what are the chances of getting an obtained mean as large as 105? The next step is to do the t test.

$$t=\frac{M_O-M_H}{S_m} \tag{26}$$

where M_O = obtained mean

M_H = hypothetical population mean of 100

$$t=\frac{105-100}{1.74}$$

$$=\frac{5.00}{1.74}=2.87$$

For this test, d.f. $=N-1=100-1=99$.

One-Tailed and Two-Tailed Tests

Obviously, you are going to reject the null hypothesis that $M_H=M_O$, but at what level? If you consult Table A–6, you will find that a t value of 2.87 occurs less than 1 time in 100 (.01) with 99 d.f. if you consider the fraction of area in both tails of the distribution. So one possibility is to state this as your level of confidence. But you might argue (and some people do) that you began with the hypothesis that the IQ in your community is higher than the population average. Would you not, therefore, be justified in considering only the area in one of the smaller tails of the distribution and rejecting the hypothesis at the .005 level of confidence.

Another way to put it is that since you predicted a mean IQ greater than 100, it is appropriate to evaluate the probability of obtaining such an outcome, ignoring the probability of obtaining a mean IQ lower than 100 by the same amount. The choice is between a one- and a two-tailed test. By this logic it is commonly argued that one-tailed tests are appropriate if your investigation involves a research hypothesis that makes predictions about the direction of differences.

TABLE A-6
Portion of a Table of *t* Values

	p*			
d.f.	.0005	.001	.01	.05
1	1,237.24	636.62	63.66	12.71
2	44.70	31.60	9.92	4.30
3	16.33	12.92	5.84	3.18
4	10.31	8.61	4.60	2.78
5	7.98	6.87	4.03	2.57
6	6.79	5.96	3.71	2.45
7	6.08	5.41	3.50	2.36
8	5.62	5.04	3.36	2.31
9	5.29	4.78	3.25	2.62
10	5.05	4.59	3.17	2.23
20	4.15	3.85	2.85	2.09
30	3.90	3.65	2.75	2.04
60	3.68	3.46	2.66	2.00
100	3.60	3.39	3.63	1.98
∞	3.48	3.29	2.58	1.96

*Both tails of the *t* distribution considered.

Although I am probably in the minority here, I favor sticking to two-tailed tests. One reason is that alternative hypotheses of some sort are always possible. Someone down the road, for example, might have the hypothesis that your community is stupider than average and offer as evidence that the fact that some of you are so worried about your intellects that you go around collecting data on the subject. Such an individual might propose to test this hypothesis with a one-tailed test by referring to the other end of the distribution. The two-tailed test considers the probability of differences in both directions. A second reason is that this conservative procedure protects you somewhat if the assumptions underlying the *t* test have not been exactly met.

Hypotheses About Mean Differences

All tests of statistical hypotheses take the same logical form: they estimate the probability of obtaining the obtained value of some statistic if the population parameter is whatever the null hypothesis states. Tests of hypotheses about differences between two means follow the same argument and ask about the probability of obtaining the obtained difference if the population is some specified number, usually zero. The steps involved are the following:

1. Obtain the mean difference of interest. There are two main situations in which this happens. I will deal first with the case where the difference is the difference between the means of two different samples.
2. Obtain the standard error of the mean by formula (25) for each sample.
3. Compute the standard error of the difference by the formula

$$S_{\text{diff}} = \sqrt{S_{M_1}^2 + S_{M_2}^2} \tag{27}$$

   where S_{M_1} and S_{M_2} are the standard errors of the mean for samples 1 and 2. This, as with all standard errors, is S of a sampling distribution, of differences in this case.
4. On the common null hypothesis that the population mean differences is zero, compute

$$t = \frac{\text{difference}}{S_{\text{diff}}} \tag{28}$$

5. Evaluate t by reference to a table of t values in terms of the appropriate degrees of freedom, rejecting the hypothesis of no difference if t is large enough.

As an example, suppose we ask whether the difference in percentages of women in the venires of the Spock judge and other Boston judges seems likely to have happened by chance. Here are the data.

	Spock Judge's Venires	Other Judges' Venires
N	9	37
Mean percentages of women	14.67	29.14
S	5.08	7.40
$\hat{S}$	5.39	7.51
S_M	1.80	1.23

Turning now to the calculations,

$$S_{\text{diff}} = \sqrt{1.80^2 + 1.23^2} = \sqrt{3.24 + 1.51} = \sqrt{4.75} = 2.18$$

$$t = \frac{29.14 - 14.67}{2.18} = \frac{14.47}{2.18} = 6.64$$

For t tests of differences, d.f. $= n_1 + n_2 - 2 = 9 + 37 - 2 = 46 - 2 = 44$ and in this case, by Table A–6, $p < .0005$.

In a way the most interesting feature of this analysis is the specific interpretation that it requires. The hypothesis under test is this. If the 9 Spock judge venires are (1) *random* samples from (2) the *same population* and if (3) the distribution of differences in terms of S_{diff} *is the t distribution*, then a t of 6.64 or larger would occur only 5 times in 10,000. On this basis, we reject the hypothesis.

But notice: The hypothesis rejected is a complex one, involving more specific hypotheses. The point to understand is that a significant t can come about if any part of this complex hypothesis is false. Usually a researcher's interest is in rejecting the part of the hypothesis which says that the two values come from the same population. This time, however, we are in the nearly unique situation of knowing that these venires *did* come from the same population, the population of selected potential jurors. For a very different set of reasons which are mathematical we can be fairly sure that the t distribution applies. This leaves as the only possibility the idea that sampling was not random.

The *t* Test for Correlated Measures

In the discussion of the wine-tasting experiments in Chapter 7, where the hypothetical study was carried out within-subjects, I presented the data shown in Table A–7. The summary materials at the bottom of the first two columns are the data we need to calculate a t value by the methods described so far.

$$t = \frac{\text{difference}}{S_{\text{diff}}} = \frac{1.0}{S_{\text{diff}}}$$

$$S_{\text{diff}} = \sqrt{(.79)^2 + (.69)^2}$$

$$= \sqrt{.62 + .48} = \sqrt{1.10} = 1.05$$

$$t = \frac{1.0}{1.05} = .95$$

TABLE A–7

	Rating		
Subject	Before Breathing	After Breathing	Difference
1	17.0	18.5	+1.5
2	17.5	18.0	+ .5
3	16.0	17.5	+1.5
4	16.5	17.0	+ .5
5	15.5	16.0	+ .5
6	15.0	15.5	+ .5
7	14.5	15.0	+ .5
8	13.0	14.0	+1.0
9	12.0	13.0	+1.0
10	9.5	12.0	+2.5
Mean	14.65	15.65	+1.0
S	2.37	2.06	.63
$\hat{S}$	2.49	2.17	.67
S_M	.79	.69	.21

This difference is not significant; in fact, far from it. You may recall, however, that the fact that every single taster gave the breathed wine a higher rating suggests a dependable effect. This, in turn, leads one to ask whether there isn't something wrong with the method of analysis just employed.

There is. If you look again at the table, you will see that the two ratings for each taster are highly correlated. The t test just applied fails to take this correlation into account. Later I shall use these same numbers to illustrate the calculation of the correlation coefficient which turns out to be +.97. The t test which takes this correlation into account does so by a different formula for S_{diff}.

$$S_{\text{diff}} = \sqrt{S_{M_1}^2 + S_{M_2}^2 - 2rS_{M_1}S_{M_2}}$$

$$= \sqrt{.79^2 + .69^2 + 2[(.97)(.79)(.69)]}$$

$$= \sqrt{1.100 - 1.057} = \sqrt{.043} = .207$$

$$= .21$$

Now the calculation of t produces a result that is quite different from the one obtained without correcting for the correlation.

$$t = \frac{1.0}{.21} = 4.76$$

The number of degrees of freedom for this test is the number of correlated pairs minus 1, or $10-1=9$. Table A-6 shows that this value falls just short of significance at $p=.001$. The probability of a Type I error is slightly over 1/1,000, which is not far from the 1/1,024, the probability that all of the differences would be in the same direction.

The Direct-Difference *t* Test

Notice that S_{diff} using the formula that corrects for the correlation in the calculations above was .21. Now look back at the last column in the table, from which these calculations were made. You will see that the value of S_M is also .21. Think what this means: The entries in that column were differences. The value of .21 is thus the *standard error of a mean difference*, or just S_{diff}. Thus in the case where the measures to be evaluated are differences,

$$S_{\text{diff}}=\frac{\hat{S}_{\text{diff}}}{\sqrt{N}} \tag{30}$$

In most cases this direct difference method of obtaining S_{diff} is much more convenient than the equivalent method which requires the calculation of r.

CORRELATION

In Chapter 8 I presented the Z score formula for r and mentioned that it would be tedious to calculate. The Z score formula is

$$r=\frac{\Sigma(Z_x \cdot Z_y)}{N} \tag{31}$$

Depending upon the way in which the data are arranged, one or the other of the following calculating formulas will be more efficient.

$$r=\frac{\dfrac{\Sigma XY}{N}-M_x M_y}{S_x S_y} \tag{32}$$

You are familiar with all the terms in this formula except ΣXY, which just indicates that you add the products of the X and Y scores for each pair as was done with Z scores in that formula. Since the denominator in the equation for r is $S_x S_y$, you can use the calculating formulas for S to arrive at other equations. One would be

$$r = \frac{\frac{\Sigma XY}{N} - M_x M_y}{\sqrt{\left(\frac{\Sigma X^2}{N} - M_x^2\right)\left(\frac{\Sigma Y^2}{N} - M_y^2\right)}} \qquad (33)$$

Formula (32) applied to the wine-testing data works out as shown in Table A–8. In formula (32), then,

$$r = \frac{\frac{2{,}340.00}{10} - (14.54 \times 15.65)}{2.37 \times 2.06}$$

$$= \frac{234.00 - 229.27}{2.37 \times 2.06} = \frac{4.73}{4.88}$$

$$= .97$$

TABLE A–8

Subject	Before Breathing, X	After Breathing, Y	XY
1	17.0	18.5	314.50
2	17.5	18.0	315.00
3	16.0	17.5	280.00
4	16.5	17.0	280.00
5	15.5	16.0	248.00
6	15.0	15.5	232.50
7	14.5	15.0	217.50
8	13.0	14.0	182.00
9	12.0	13.0	156.00
10	9.5	12.0	114.00
Total	146.5	156.5	2,340.00
Mean	14.65	15.65	234.00
S	2.37	2.06	

The Rank-Difference Method

Strictly speaking, r is appropriate only when the scales of measurement are interval or ratio scales. Where ordinal scales are involved, a formula for a different correlation coefficient (rho) based on ranks is more in order.

$$\text{rho} = 1 - \frac{6 \Sigma D^2}{N(N^2 - 1)} \qquad (34)$$

where $D=$ the difference in the ranks of the X and Y scores and N is the number of paired scores.

On the walls of the study of someone I know, there are five paintings. Here are their sizes and what they cost:

Size (in.)	Cost
10×12	$ 10
18×24	80
26×20	15
30×36	475
52×80	900

If you rank these values from least to most, you get the following arrangement:

Rank in Size	Rank in Cost	D	D^2
1	1	0	0
2	3	1	1
3	2	1	1
4	4	0	0

The value of rho, thus, is

$$\text{rho} = 1 - \frac{6\Sigma D^2}{N(N^2-1)} = 1 - \frac{6\times 2}{5(24)} = 1 - \frac{12}{120}$$
$$= 1 - .1 = +.90$$

ANALYSIS OF VARIANCE

A deeper understanding of ANOVA requires a knowledge of two points: (1) that variance can be analyzed into components and (2) that estimates of population variance can be made on the basis of these components. In Chapter 10 I tried to put these ideas across in a nontechnical way which I, at least, find only somewhat satisfactory. In this section I will present more formal proofs of some of the important points. I suspect that some of you, like me, will be a bit more comfortable after such a demonstration.

Definitions

1. n = the number of individuals in each of several groups. I shall deal only with the case where n is equal for all groups
2. k = the number of groups; thus
3. $kn = N$, the total number of individuals in all groups
4. X = any score
5. M = the mean for an individual group
6. GM = the general mean, the mean of all N scores (also the mean of the group means)
7. d_T = the total (T) deviation (d) of a score from the general mean = $X - \text{GM}$
8. d_{WG} = the deviation of a score from its own sample mean = $X - M$
9. d_{BG} = the difference between a sample mean and the general mean = $M - \text{GM}$

Proof That $d_T = d_{WG} + d_{BG}$

1. $d_T = X - \text{GM}$, by definition 7
2. $d_{WG} + d_{BG} = X - M + M - \text{GM}$ by definitions 8 and 9 $= X - \text{GM}$
3. $d_T = d_{WG} + d_{BG}$

Proof That $\Sigma d_T^2 = \Sigma d_{WG}^2 + \Sigma d_{BG}^2$

1. $d_T = d_{WG} + d_{BG}$
2. $d_T^2 = (d_{WG} + d_{BG})^2$
   $= d_{WG}^2 + 2(d_{WG} \cdot d_{BG}) + d_{BG}^2$
3. Adding for any individual group,

$$\Sigma d_T^2 = \Sigma d_{WG}^2 + 2(\Sigma d_{WG} \cdot \Sigma d_{WG}) + d_{BG}^2$$

4. Since, as you know, $d_{WG} = 0$ (actually, Σd_{BG} also $= 0$), the middle term drops out and

$$\Sigma d_T^2 = \Sigma d_{WG}^2 + \Sigma d_{BG}^2 \tag{35}$$

5. Alternatively, since $d_{BG} = M - \text{GM}$, a constant in any group

$$\Sigma d_T^2 = \Sigma d_{WG}^2 + n d_{BG}^2 \tag{36}$$

6. In the language of analysis of variance, it is customary to refer to any Σd^2 as a "sum of squares" and to use the abbreviation SS with appropriate subscripts instead of the expressions above. Thus

$$\text{SS}_T = \text{SS}_{WG} + \text{SS}_{BG} \tag{37}$$

Obtaining Components of Variance

1. From formula (36) for any single group,

$$\Sigma d_T^2 = \Sigma d_{WG}^2 + n d_{BG}^2$$

2. Adding these terms for all k groups,

$$\Sigma(\Sigma d_T^2) = \Sigma(\Sigma d_{WG}^2) + n\Sigma d_{BG}^2 \tag{38}$$

It will be important later to remember that SS_T breaks down into components in exactly this way.

3. Understanding that Σd_T^2 is the sum across all groups and dividing by $N = kn$,

$$\frac{\Sigma d_T^2}{N} = \frac{\Sigma(\Sigma d_{WG}^2)}{kn} + \frac{\not{n}\Sigma d_{BG}^2}{k\not{n}} \tag{39}$$

Alternatively,

$$\frac{\Sigma d_T^2}{N} = \frac{1}{k}\Sigma\left(\frac{\Sigma d_{WG}^2}{n}\right) + \frac{\Sigma d_{BG}^2}{k} \tag{40}$$

You should recognize the first right-hand term as the mean of k within-group variances. The expression directs you to add these variances $\Sigma(\Sigma d_{WG}^2/n)$ and divide by k. The second right-hand term is the variance of the means. These are the two variances into which I divided total variance in Chapter 10 by using numerical examples.

From Variance Back to Σd^2

In Chapter 10 I multiplied S^2 by N to obtain Σd^2. Formula (39) begins to show why this can be done. For total variance, since

$$S_T^2 = \frac{\Sigma d_T^2}{N} \quad \text{and} \quad NS_T^2 = \Sigma d_T^2 \tag{41}$$

For variance within groups, since

$$S_{WG}^2 = \frac{\Sigma(\Sigma d_{WG}^2)}{kn}$$

$$= \frac{\Sigma d_{WG}^2}{N}$$

understanding that the summation is now across all groups,

$$NS_{WG}^2 = \Sigma d_{WG}^2 \tag{42}$$

For variance between groups, before canceling n in formula (39),

$$S_{BG}^2 = \frac{n \Sigma d_{BG}^2}{nk} = \frac{n \Sigma d_{BG}^2}{N}$$

$$NS_{BG}^2 = n \Sigma d_{BG}^2 \tag{43}$$

Estimating Population Variance

The components of total variance previously obtained do not provide unbiased estimates of population variance. The obtaining of such estimates is straightforward, however.

The Within-Groups Estimate. For any one of the k groups, the best estimate of population variance is

$$S_{WG}^2 = \frac{\Sigma d_{WG}^2}{n-1}$$

The average of these estimates is still a better estimate:

$$\begin{aligned} S_{WG}^2 &= \frac{1}{k} \Sigma \left(\frac{\Sigma d_{WG}^2}{n-1} \right) \\ &= \frac{\Sigma (\Sigma d_{WG}^2)}{k(n-1)} \\ &= \frac{\Sigma (\Sigma d_{WG}^2)}{kn-k} \\ &= \frac{SS_{WG}}{N-k} \end{aligned} \tag{44}$$

The Between-Group Estimate. This estimate depends upon the fact that the between-groups sum of squares makes it possible to estimate S_M^2. For each of the k samples, an estimate of population variance can be obtained this way.

$$S_M^2 = \frac{\hat{S}^2}{n} \quad \text{and} \quad nS_M^2 = \hat{S}^2$$

The best estimate of the variation of the population of means from the k means available is

$$S_M^2 = \frac{\sum d_{BG}^2}{k-1}$$

Thus, multiplying by n,

$$nS_M^2 = \hat{S}_{BG}^2 = \frac{n\sum d_{BG}^2}{k-1}$$

$$\hat{S}_{BG}^2 = \frac{\text{SS}_{BG}}{k-1} \qquad (45)$$

The Partitioning of Degrees of Freedom. The derivations just carried out also divide up the available degrees of freedom appropriately. For all N measures there are $N-1$ d.f.

We have assigned $N-k$ d.f. to the within-groups estimate of population variance and $k-1$ d.f. to the between-groups estimate. Thus

$$(N-k)+(k-1)=N-k+k-1=N-1$$

Calculating Formulas

In a moment I shall analyze the results of a 4×4 factorial experiment with the methods of ANOVA. As a preliminary, I believe that it will provide helpful orientation if I give you an overview of the calculational procedures, presenting the materials in terms of sums of squares.

Formula (37) tells you that

$$\text{SS}_T = \text{SS}_{WG} + \text{SS}_{BG}$$

This means, of course, that

$$\text{SS}_{WG} = \text{SS}_T - \text{SS}_{BG} \qquad (46)$$

This fact becomes very useful because SS_{WG} would be a nuisance to calculate. The fact that it can be obtained by subtraction saves a great deal of time.

The actual experiment I will analyze is a 4×4 factorial experiment with four values of two different conditions and 16 (4×4) groups. Such an experiment, as I showed in Chapter 10, allows you to analyze the variance among these 16 groups (SS_{BG}) further.

As you also saw in Chapter 10, factorial experiments with several values of only two variables can be laid out in a matrix where the columns (C) represent the values of one variable and the rows (R) represent the values of the other. Using R and C to identify the sums of squares associated with these two variables, the between-groups sum of squares can be analyzed as implied in this equation:

$$SS_{BG} = SS_R + SS_C + SS_{R \times C} \tag{47}$$

where $SS_{R \times C}$ is a sum of squares for the interaction between R and C. Again, the additivity of these sums of squares solves some computational problems, this time those involved in computing $SS_{R \times C}$ because

$$SS_{R \times C} = SS_{BG} - SS_R - SS_C \tag{48}$$

Interim Summary. Putting all of this together, we can say that

$$SS_T = SS_{WG} + SS_R + SS_C + SS_{R \times C} \tag{49}$$

In actual calculations, the procedure will be:

1. Calculate SS_T by formula (50).
2. Calculate SS_{BG} by formula (51).
3. Calculate $SS_{WG} = SS_T - SS_{BG}$ (formula 46).
4. Calculate SS_C by a variation of formula (51).
5. Calculate SS_R by the same variation of formula (51).
6. Calculate $SS_{R \times C} = SS_{BG} - SS_R - SS_C$ (formula 48).
7. Compute Fs by formula (52).

The several formulas mentioned are all closely related to formula (11), a calculating formula for variance:

$$S^2 = \frac{\Sigma X^2 - (\Sigma X)^2 / N}{N} \tag{11}$$

$$NS^2 = \Sigma X^2 - \frac{(\Sigma X)^2}{N} = \Sigma d^2 = SS$$

With these preliminaries out of the way, I can now proceed with an example, planning to develop some specifics as I go along.

A DOWN-TO-EARTH EXAMPLE

Much of the pioneering work in statistics was done by researchers in the field of agriculture. My final set of data is from such a study.[2] The study was a greenhouse investigation of the yield of wheat grown in soils treated in different ways. The experiment was a 4×4 factorial in which four chemical treatments of soil (1, 2, 3 and 4) were crossed with four different types of fertilizer (A, B, C, and D). This design yields 16 experimental conditions (4×4). Three seperate pots of wheat were planted for each of the 16 conditions. These pots were placed at random around the greenhouse. This is important in order to assure that the outcomes were not the result of favorable or unfavorable placements.

The data are shown in Table A–9. I wish to call your attention to the fact that totals rather than means are entered as summarizing statistics. This is because the computation makes use of those values.

TABLE A–9
Yields of Wheat in 48 Pots

Fertilizer Treatment	Pot	Chemical Treatment 1	2	3	4	Total 12 Pots
A	1	21.4	20.9	19.6	17.6	
	2	21.2	20.3	18.8	16.6	
	3	20.1	19.8	16.4	17.5	
	Total	62.7	61.0	54.8	51.7	230.2
B	1	12.0	13.6	13.0	13.3	
	2	14.2	13.3	13.7	14.0	
	3	12.1	11.6	12.0	13.9	
	Total	38.3	38.5	38.7	41.2	156.7
C	1	13.5	14.0	12.9	12.4	
	2	11.9	15.6	12.9	13.7	
	3	13.4	13.8	12.1	13.0	
	Total	38.8	43.4	38.9	39.1	160.2
D	1	12.8	14.1	14.2	12.0	
	2	13.8	13.2	13.6	14.6	
	3	13.7	15.3	13.3	14.0	
	Total	40.3	42.6	41.1	40.6	164.6
	Total 12 pots	180.1	185.5	173.5	172.6	711.7

[2] G. W. Snedecor, *Statistical Methods*, 3rd ed. (Ames, Iowa: The Iowa State College Press, 1940).

The analysis begins with the calculation of SS_T by the formula derivable, as mentioned earlier, from formula (11):

$$SS_T = \Sigma X^2 - \frac{(\Sigma X)^2}{N} \qquad (50)$$

The last term in this formula is a *correction factor*. As was shown in formula (10), this correction factor

$$\frac{(\Sigma X)^2}{N} = NM^2$$

In effect, what it does in formulas for SS is to subtract M from each of the scores to be squared, as is required in the expression $\Sigma d^2 = SS$. The correction factor for these data is

$$\frac{(\Sigma X)^2}{N} = \frac{(711.7)^2}{48} = \frac{506{,}516.89}{48} = 10{,}552.44$$

Now, applying formula (50),

$$SS_T = (21.4)^2 + (21.2)^2 + \cdots + (14.6)^2 + (14.0)^2 - 10{,}552.44$$
$$= 367.15$$

The next step is to compute the sum of squares between groups by the formula

$$SS_{BG} = \sum_{1}^{k} \frac{(\Sigma X)_i^2}{n} - \frac{(\Sigma X)^2}{N} \qquad (51)$$

where $(\Sigma X)_i =$ the total for each individual (i) group. The formula directs you to square each such total, divide by the number of cases in the group (n), add up the $k = 16$ quotients, and subtract the correction factor. In the example

$$SS_{BG} = \frac{(62.7)^2}{3} + \frac{(38.3)^2}{3} + \cdots + \frac{(40.6)^2}{3} - 10{,}552.44$$
$$= 10{,}893.31 - 10{,}552.44 = 340.87$$

With SS_T and SS_{BG} available, SS_{WG} can be obtained by subtraction:

$$SS_{WG} = 367.15 - 340.87 = 26.28$$

If this were a multivalent experiment with 16 different values of a simple variable, this is as far as the obtaining of sums of squares would go. You would estimate population variance from these two sums of squares—in effect applying formulas (44) and (45) by dividing by the appropriate numbers of d.f. and do the F test:

$$\hat{S}^2_{BG} = \frac{367.15}{15} = 24.48$$

$$\hat{S}^2_{WG} = \frac{26.28}{32} = .82$$

$$F = \frac{\hat{S}^2_{BG}}{\hat{S}^2_{WG}} \tag{52}$$

$$= \frac{24.48}{.82} = 29.85$$

In our example there were 16 conditions and 3 pots of wheat in each. Hence $N=48$ and the total number of d.f. $=48-1=47$. The degrees of freedom for the $k=16$ groups are $k-1=15$ and d.f. within groups (error) are $N-k=48-16=32$. For 15 and 32 d.f., the F just calculated is highly significant but of no real interest, because in this case between-groups variance is a mish-mash made up of variances due to chemical treatment, fertilizer treatment, and an interaction. The results become meaningful only when these effects are examined separately.

The sums of squares for chemical treatments and fertilizer treatments are calculated by the same general formula as the between-groups sum of squares (51).

$$SS_{BG} = \sum_{1}^{k} \frac{(\Sigma X)_i^2}{n} - \frac{(\Sigma X)^2}{N}$$

This time, however, the sum of squares for chemical treatments would be obtained by "collapsing across fertilizer treatments," that is, by adding all the scores for each of the different chemical treatments together, ignoring the different fertilizer treatments. Because of this procedure, n in the formula becomes 12 for each of 4 sets of scores (columns):

$$SS_{\substack{\text{chemical}\\\text{treatment}}} = \frac{(180.1)^2}{12} + \frac{(185.5)^2}{12} + \frac{(173.5)^2}{12} + \frac{172.6}{12} - 10{,}552.44$$

$$= 10{,}561.61 - 10{,}552.44 = 9.17$$

The calculation follows the same pattern for fertilizer treatment:

$$SS_{\substack{\text{fertilizer}\\\text{treatment}}} = \frac{(230.2)^2 + (156.7)^2 + (160.2)^2 + (164.6)^2}{12} - 10{,}552.44$$

$$= 10{,}858.68 - 10{,}552.44 = 306.24$$

We have left to figure only the sum of squares for interaction, which can be obtained by the subtraction indicated in formula (48).

$$SS_{R \times C} = SS_{BG} - SS_R - SS_C$$

$$SS_{\text{interaction}} = 340.87 - 9.17 - 306.24 = 25.46$$

The fruits of our labors can be summarized in an ANOVA table.

Source of Variation	d.f.	Sum of Squares	$\hat{s}^2$	F
Chemical treatment	3	9.17	3.06	3.73*
Fertilizer treatment	3	306.24	102.08	124.49†
Interaction	9	25.46	2.83	3.45*
Within groups (error)	32	26.28	.82	
Total	47	367.15		

*$p < .05$.
†$p < .01$.

In this table successive columns give the usual information. The chief point worth calling your attention to may be the entries for d.f. Just to make sure that it is clear, let me mention that four chemical treatments yield 3 d.f. and four fertilizer treatments do the same. There are 9 d.f. for the interaction (3×3). As we have seen before, these add up to the d.f. between groups: $9+6+6=15=k-1$. As you also saw earlier, there are $N-k=48-16=32$ d.f. for the error term.

The table shows that the two main effects are significant, chemical treatment at the .05 level and fertilizer treatment at well beyond the .01 level. The significant interaction means that the effectiveness of the different fertilizer treatments varied with the particular chemical treatment. Overall fertilizer treatment A was most effective, but it worked its best with chemical treatment 1 and worst with chemical treatment 4. For fertilizer treatment B, the results were exactly opposite. Fertilizer treatments C and D were most effective with chemical treatment 2. The general meaning of an interaction is that the effects of one variable are different in a way that depends upon the value of another variable. In terms of a table such as the one used, another way to put it is that column-to-column differences differ from row to row.

End Notes

This final section of the book does several things: (1) in some cases materials that would have interrupted the argument of the main text appear here; (2) In other cases I will introduce new materials that seem to me to offer useful extensions of what I have covered earlier; and (3) Finally, for each chapter I will present a few problems for your consideration. Solutions to the problems appear at the end of the notes for each chapter.

CHAPTER 1. THE NATURE OF STATISTICS

Problems

1. A metropolitan newspaper proclaimed in one of its headlines that the average cost of a home in one of its suburban areas was $75,000. The Board of Realtors in the suburbs responded, claiming that the value was more like $60,000. How could such a disagreement come about?

2. Back in the days of 78-rpm records, there was a comedian whose albums were a source of scandalous delight to the youths of those days. I no longer remember his name, but I do remember a bit he did that I can use as a basis for asking some questions.

The scene is an Old Men's Club in London where the criterion for admission is age. The occasion is the annual meeting of the club, and most of the members are present. Some are 99, some are 100, some are 101, and some are even older. Everyone is very excited and dancing around the chair because the word has just come in that one of their most senior members, old Lord Putney, has married a beautiful young American heiress with pots and pots of gold. Everyone that is except the club cynic, Lord Sackett. Asked about the news, all he has to say is, "Wait and see."

And now it is a year later. Again it is the annual club meeting and again most of the membership is there. Some are 100, some are 101, some are 102, and some are even older. Once more the members are all excited and dancing around the chair because now the news is that the beautiful young

American heiress with pots of gold has given birth to a son. And what does Lord Sackett think of that? "It reminds me of Africa," he answers.

"Africa?" they respond in feeble chorus, "How so?"

"Only last year," replies Lord Sackett, "I was in Africa, on safari, hunting lions—even at my advanced age. The luck was beastly (no pun intended) and I grew careless.

"One night I went unarmed into the bush and out of a thicket there came rushing at me a huge ferocious lion. Without a weapon I didn't know what to do. But reflexively, I suppose, I raised an imaginary gun and shouted, 'Bang! Bang! Bang!'. And lo and behold, the lion fell dead at my feet.

"And then I turned. And there at my side was a much much younger man with a smoking rifle in his hands."

Problem: You are, let us say, Lord Sackett. What is your research hypothesis? What is the null hypothesis? How could the null hypothesis be tested? Can you use these data to explain the concept of level of confidence?

Outside Reading

You know by now, if you have read this book, or will find out, if you are about to, that it is not so much a statistics book as it is a book about statistics. Some people find my presentation disconcerting because I do so little with formulas and calculations. If you discover that you are one of those people, you may want to find another, more traditional book to study along with this one. These standard textbooks will have a more formal development of topics and a greater number of problems for you to do. I would recommend that you go to a secondhand bookstore and find the cheapest text you can. Because of the kinds of materials covered in this book, a text on psychological statistics might be most useful.

Three other fairly nontechnical books that you might enjoy are these:

Huff, D. *How To Lie with Statistics*. New York: W. W. Norton, 1954.

Schutte, J. G. *Everything You Always Wanted To Know About Elementary Statistics (But Were Afraid To Ask)*. Englewood Cliffs, N.J.: Prentice-Hall, 1977.

Wallis, W. A., and H. V. Roberts, *The Nature of Statistics*. New York: Free Press, 1956.

Solutions to Problems

1. As housing prices spiral upward, the statistics on costs are becoming a maze of contradictions. In the particular instance mentioned, what was involved was a very simple matter of operational definition: $75,000 were asking prices; $60,000 were selling prices. Other actual statistics I have seen in recent months put the increase in costs of private homes anywhere between 1.4% and 13%. To get the lowest percentage, it is necessary to include (a) rural housing, (b) vacant lots, (c) condominiums, and (d) houses in such delapidated condition that they were purchased just for the value of the land they occupied. To get the higher figure, all of these are excluded and the increase is figured for just the most desirable urban dwellings.

2. Research hypothesis: Some much, much younger man was the father of the child born by the beautiful young American heiress with pots and pots of gold. Null hypothesis: The much, much younger man was *not* the father. The only realistic test of the hypothesis is unrealistic. It would require the collection of data on a large number of couples as mismatched as Lord Putney and his bride. Ideally, it would also require separation of these couples from the company of younger men. The data would be the number of births in a reasonable period of time. Suppose there were 1 in 1,000 cases, .1%. On this basis, you reject the null hypothesis at the .1% level of confidence. The result could happen even if the null hypothesis is true, but the chances are 1 in 1,000.

CHAPTER 2. PICTURES OF DATA

Problems

1. Data on the birth dates and death dates are available for 1,251 well-known people [D. P. Phillips, "Death and birthday: an unexpected relationship," in J. M. Tanur, F. Mosteller, W. H. Kruskal, R. F. Link, R. S. Pieters, and G. R. Rising (eds.), *Statistics: A Guide to the Unknown* (San Francisco: Holden-Day, 1972)]. When these data are presented in terms of the number of months before or after their month of birth that these people died, the results are as shown in the table below.

Months Before						Birth Month	Months After				
6	5	4	3	2	1	0	1	2	3	4	5
90	100	87	96	101	86	119	118	121	114	113	106

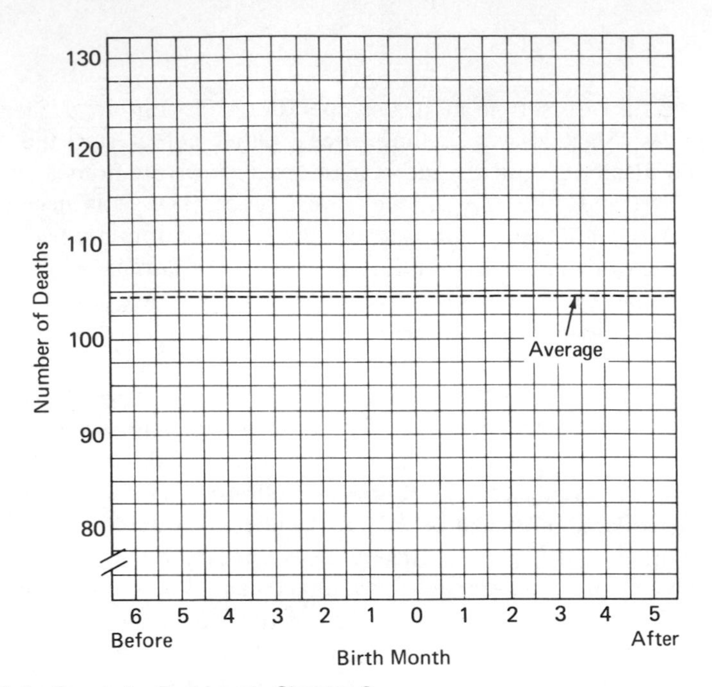

Figure N–1 Graph for Problem 1, Chapter 2.

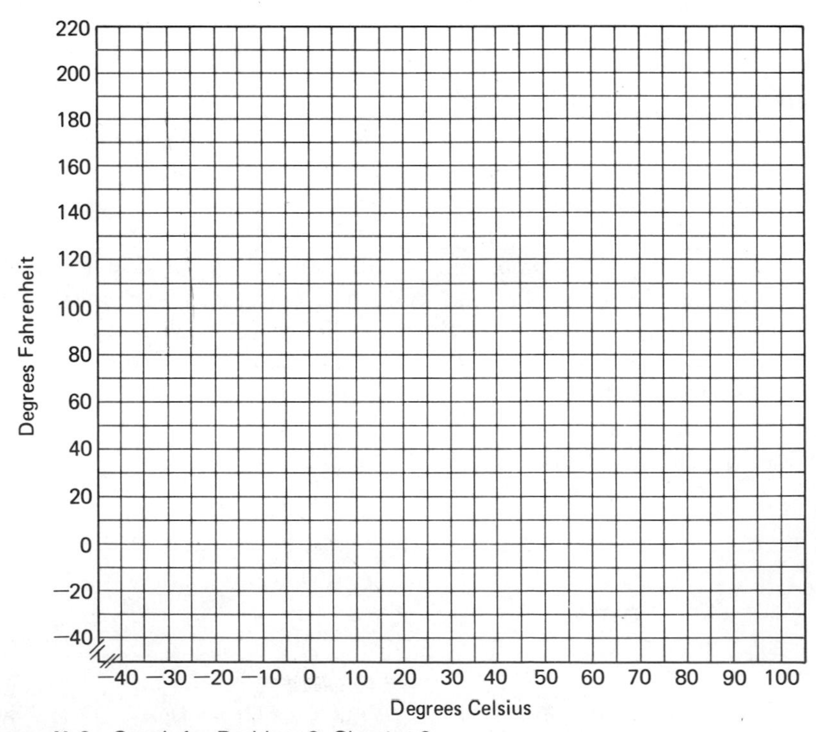

Figure N–2 Graph for Problem 2, Chapter 2.

Plot these data on the coordinates provided (Figure N–1). What does the resulting graph appear to indicate? The "average" line in the graph is set at the level $1{,}251 \div 12 = 104.25$, 1,251 being the total number of deaths.

2. The formula for translating degrees centigrade to Fahrenheit is

$$F^\circ = \tfrac{9}{5}C + 32$$

Going from Fahrenheit to centigrade, the formula is

$$C^\circ = \tfrac{5}{9}(F^\circ - 32)$$

The equations yield the following equivalencies.

°C	°F
−40	−40
−30	−22
−20	−4
−10	14
0	32
10	50
20	68
30	86
40	104
50	122
60	140
70	158
80	176
90	194
100	212

Plot these data on the coordinates provided (Figure N–2). How do you describe the resulting function? Can you justify this visual impression with arithmetic?

Answers to Problems

1. The plot of number of deaths against months before and after the birth month (see Figure N–3) shows that people (at least famous people) tend to postpone death until after a birthday.

2. Plot of Fahrenheit degrees against Celsius degrees is linear. As Figure N–4 shows, every change of 10° Celsius is associated with a change of 18° Fahrenheit. These equal changes define a linear function.

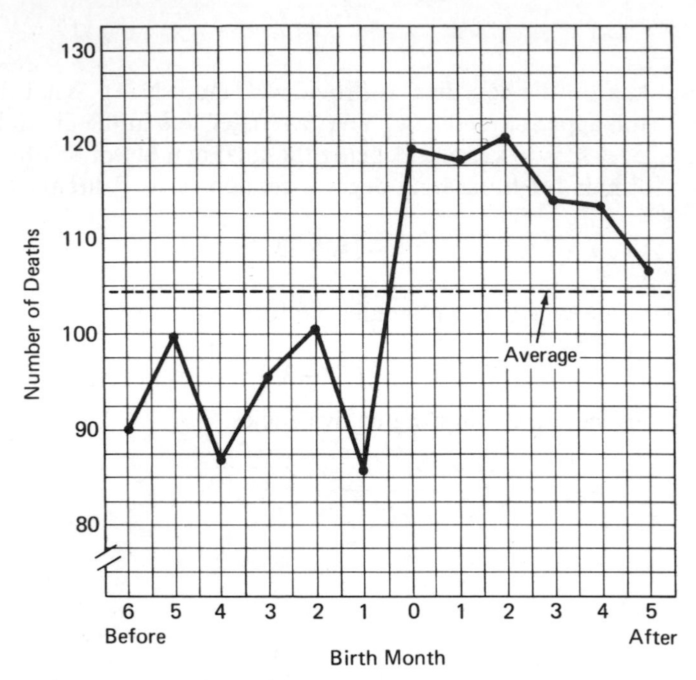

Figure N–3 Solution to Problem 1, Chapter 2.

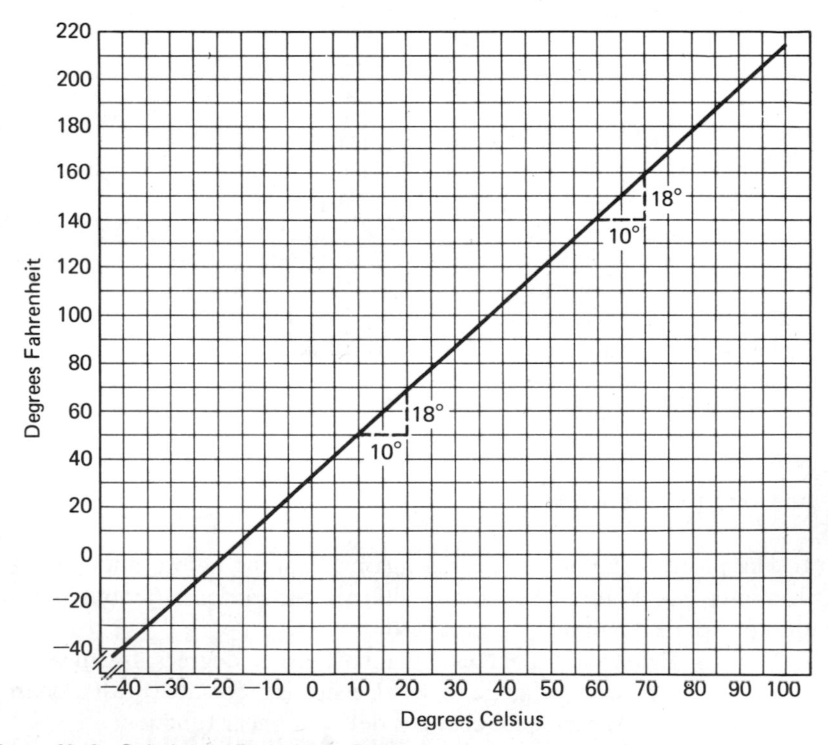

Figure N–4 Solution to Problem 2, Chapter 2.

CHAPTER 3. FREQUENCY DISTRIBUTIONS

Problems

J. E. Freund and F. J. Williams, *Elementary Business Statistics: The Modern Approach* (Englewood Cliffs, N.J.: Prentice-Hall, 1972) present the following data as the number of minutes 100 customers occupied their seats in a cafeteria.

29	67	34	39	23	66	24	37	45	58
51	37	45	26	41	55	27	96	22	43
73	48	63	37	19	31	38	68	22	35
31	58	35	82	28	35	44	40	41	34
15	31	34	56	45	27	54	46	62	29
51	31	56	43	39	35	23	28	45	48
47	41	34	47	30	54	49	34	53	61
82	45	26	35	67	73	30	16	52	35
46	40	41	56	37	51	33	92	70	63
72	35	62	28	38	61	33	49	59	35

1. Make a tally of these data using these class intervals 10–19, 20–29, 30–39, 40–49, 50–59, 60–69, 70–79, 80–89, 90–99.

2. Using the coordinates provided (Figure N–5), construct a bar graph to represent these data.

3. Convert the bar graph into a frequency polygon defining the scores as the midpoints of the class intervals: 14.5, 24.5, etc.

4. Which of the following designations apply to these distributions? Symmetrical, positively skewed, negatively skewed, J curve.

5. Convert the data to a cumulative frequency distribution (Figure N–6), using the upper limit of the class intervals as scores on the abscissa. This is to give you the number of scores exceeded by a given score.

Frequencies of Trivia

The data from which Figure 3–2 was constructed came from the following source: F. L. Worth, *The Trivia Encyclopedia* (Los Angeles: Brooke House, 1974). The items contributing to the figure are outlined below.

Zero things: zero frequency.

One thing: zero frequency.

Two things: frequency of 1. Two thieves crucified with Jesus.

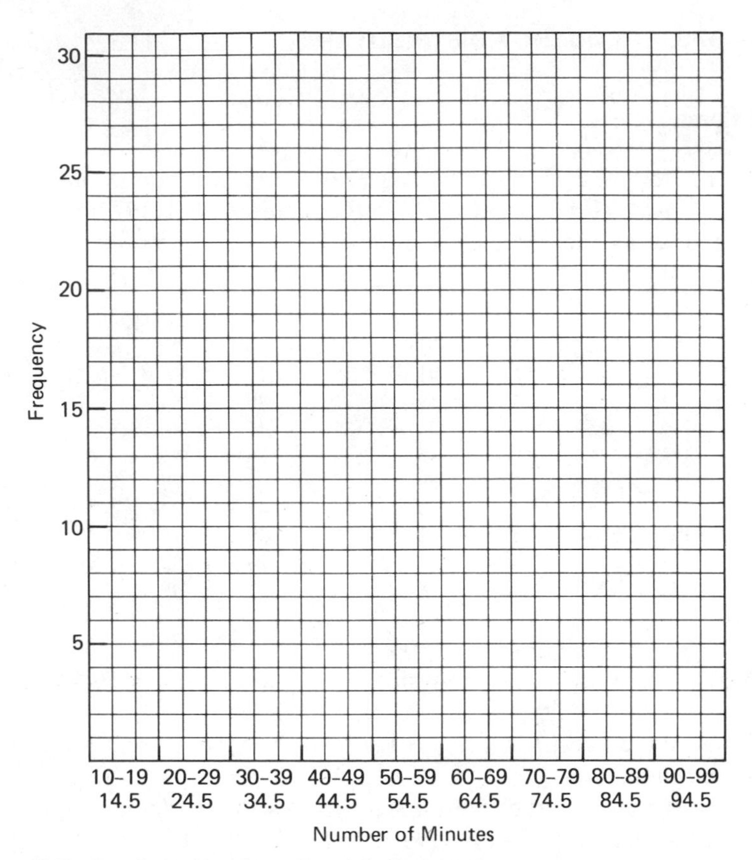

Figure N–5 Graph for Problems 2 and 3, Chapter 3.

Three things: frequency of 15. 3 Bs (Bach, Beethoven, Brahms), 3 Caballeros (Ponchito, Jose Carioca, and Donald Duck), 3 cardinal virtues (Faith, Hope, and Charity), 3 fates (Lachesis, Clotho, Atropos), 3 furies (Alecto, Magaera, Tisiphone), 3 good fairies (Flora, Fauna, and Merryweather in Walt Disney's *Sleeping Beauty*), 3 graces (Aglaia, Thalia, Euphrosyne—daughters of Zeus and Eurynome), 3 little maids (Yum Yum, Peep-Bo, Pitti-Sing, in Gilbert and Sullivan's *Mikado*), 3 men in a tub (butcher, baker, and candlestick maker), 3 musketeers (Athos, Porthos, Aramis), 3 musketeers motto (All for one, one for all), 3 r's (reading, "riting," "rithmetic"), 3 stooges, 3 wise men, 3 wise monkeys (see no evil, hear no evil, speak no evil).

Four things: frequency of 7. 4 chaplains aboard USS Dorchester who gave up their life jackets to other soldiers on 3/3/43, 4 dimensions (length, width, depth, time), 4 evangelists (Mathew, Mark, Luke, John), 4 freedoms (of speech, of religion, from want, from fear—F. D. Roosevelt, 1/6/41), 4

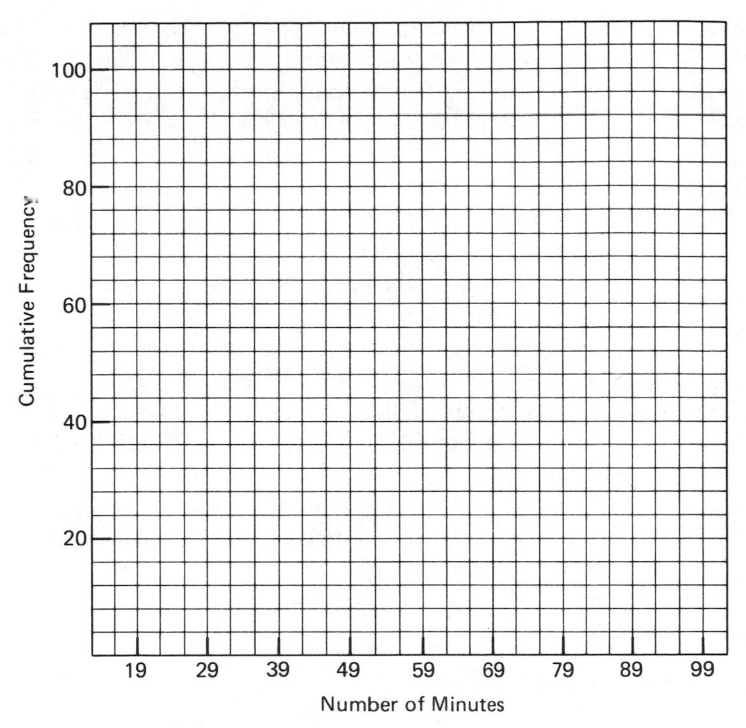

Figure N–6 Graph for Problem 5, Chapter 3.

horsemen of Notre Dame—the backfield of Stuhldreher, Miller, Crowley, and Layden), 4 humors (black bile, yellow bile, phlegm, and blood), 4 horsemen of the apocalypse (conquest, famine, pestilence, and death).

Five things: frequency of 6. 5 books of Moses (Genesis, Exodus, Leviticus, Numbers, Deuteronomy), 5 civilized tribes (Cherokee, Choctaw, Chickasaw, Creek, Seminole), Five Little Peppers and How They Grew (Ben, Phronsie, Polly, Joel, and Davie in book by Margaret Sidney, 1881), 5 nations (the Iroquois League: Seneca, Mohawk, Oneida, Onondaga, Cayuga), Five pennies (Red Nichols band), Five Points (town of radio series "The Guiding Light"), 5 rivers of Hades (Acheron = woe, Cocytus = lamentation, Lethe = oblivion, Phlegethon = fire, Styx = hate), 5 years (duration of "Star Trek" mission), 5 Ws (who, what, where, when, why).

Six things: frequency of 3. 6 sides to a snowflake, 6 players on a hockey team, 6 questions asked by a good reporter (who, what, where, when, why, and how).

Seven things: frequency of 16. 7 heroes who fought against Thebes (Adrastus, Polynices, Tydeus, Parthenopaeus, Amphiaraus, Capaneous,

Hippomedon), 7 voyages of Sinbad the sailor, 7 ages of man (infant, schoolboy, lover, soldier, justice, retirement, second childhood), 7 deadly sins (pride, avarice, wrath, envy, gluttony, sloth, lust), 7 destroyers ran aground in the Santa Barbara channel, 7/8/23), 7 dwarfs in Snow White (Bashful, Doc, Dopey, Grumpy, Happy, Sleepy, Sneezy), 7 Hills of Rome, 7 little Foys, 7 mules (linemen for the 4 horsemen (Huntsinger, Collins, Bach, Miller, Kizer, Weiber, Walsh), 7 seas (Antarctic, Arctic, North Atlantic, South Atlantic, Indian, North Pacific, South Pacific), 7 sisters (Women's Ivy League colleges: Barnard, Bryn Mawr, Mount Holyoke, Radcliffe, Smith, Vassar, Wellesley), 7 sisters (the constellation Pliades), 7 virtues (faith, hope, charity, fortitude, justice, prudence, temperance), 7 wonders of the Ancient World (Collosus of Rhodes, Egyptian pyramids, Hanging Gardens of Babylon, lighthouse at Alexandria, mausoleum at Halicarnassus, statue of Zeus at Olympia, Temple of Diana at Ephesus), 7 wonders of the Middle Ages (catacombs of Alexandria, Colliseum of Rome, Great Wall of China, Leaning Tower of Pisa, St. Sophia's mosque at Constantinople, Porcelain Tower of Peking, Stonehenge in England), 7 works of mercy (bury the dead, clothe the naked, feed the hungry, give drink to the thirsty, house the homeless, tend the sick, visit the orphaned and afflicted).

Eight things: frequency of zero.

Nine things: frequency of 1. 9 muses (Clio—history, Melpomene—tragedy, Thalia—comedy, Calliope—epic poetry, Urania—astronomy, Euterpe—lyric poetry, Terpischore—dance, Polymnia—song, Erato—love poetry).

Ten things: frequency of 1. 10 plagues of Egypt [water turns to blood, frogs, lice, flies, cattle murrain (sickness), boils, hail and fire, locusts, darkness, slaying of firstborn].

Answers to True–False Test

1. *True*, although social expectations may encourage redheads to be slightly more hot-tempered.

2. *False*; the data are fairly clear in showing the nonreacting participants who may even witness murder without intervening stand by in groups. A single person is more apt to try to help.

3. *True*; known metabolic and neurological factors account for, perhaps, one-third of the cases.

4. *False*; there is now some evidence for a slight tendency toward abnormality in very bright people but "closely related" is far too strong.

5. *False*, unless you put in that category the fact that many of them turn out to be preachers.

6. *True*; in general there is a positive relationship between college grades and later financial success.

7. *False*; data are mixed, but it seems more likely that TV violence provides a model for violent behavior and increases it.

8. *True*; the opposite is an unfortunate myth.

9. *True*; there is actually a very slight positive relationship between physical and mental strength.

10. *True*; for example, at the physiological level we all have both male and female hormones.

11. *False*; such letters are essentially worthless. Tests, whatever their weaknesses, are somewhat better.

12. *False*, although a handicapped person may *learn* to use other talents.

13. *True*, but even a 1% error can make a poll wrong in a close vote.

14. *True*; accurate judgments require a knowledge of the *situation* giving rise to the emotional behavior.

15. *True*; the key word is *lights*. Paints produce a muddy brown.

16. *False*; the positive demonstrations have serious methodological flaws.

17. *True*, a part of the problem being that we all have experiences that might have serious consequences. Usually, mental illness culminates a history of maladjustment.

18. *False*; we now know that genetic factors are important and that it was a sexist error to blame it on mothers in any case.

19. *False*; the methods are 200 percent American. Thorndike and Skinner are the important people in the history of the laboratory methods. They never advocated ideological application of the methods, however.

20. *True*; on the basis of experimental evidence, a year or two of preschool experience will do it.

21. *False*; they tend to be only children and firstborns.

22. *False*; likes tend to attract.

23. *True*; white rats are albinos and like human albinos have only black–white–gray vision and poor acuity.

24. *False*; if anything, they are insensitive and unable to discriminate hunger pangs from anxiety and general arousal.

25. *True*; hypnosis physiologically is more like the waking state.

26. *False*; the earliest smiles are reflexes elicitable by any facelike stimulus.

27. *False*; the common belief in the truth of this statement is a tragic misconception. Lives might have been saved if people had known that the person who threatens suicide is serious.

28. *True*; there is much evidence to support the point.

29. *True*; the probability of the death of a person is below chance expectation for the few months before his or her birthday (at the very lowest in the month immediately prior) and above chance for a few months after.

30. *True*, especially for materials learned by rote memory and with little understanding.

Solutions to Problems

1. *Tally of data*

10–19	1 1 1
20–29	~~1111~~ ~~1111~~ 1 1 1 1
30–39	~~1111~~ ~~1111~~ ~~1111~~ ~~1111~~ ~~1111~~ 1 1 1 1
40–49	~~1111~~ ~~1111~~ ~~1111~~ ~~1111~~ 1 1
50–59	~~1111~~ ~~1111~~ 1 1 1 1
60–69	~~1111~~ ~~1111~~
70–79	1 1 1 1
80–89	1 1
90–99	1 1

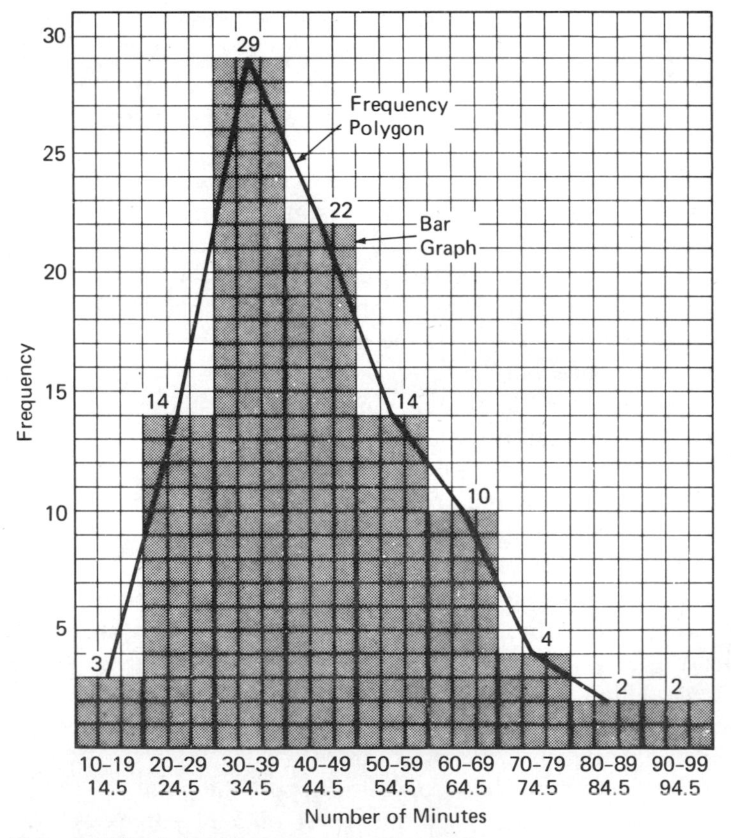

Figure N–7 Solution to Problems 2 and 3, Chapter 3.

2, 3, 4. *Graphic representations.* The two constructions in Figure N–7 show that the graph is positively skewed. The tail is toward the larger numbers.

5. Cumulative frequency distribution. See Figure N–8.

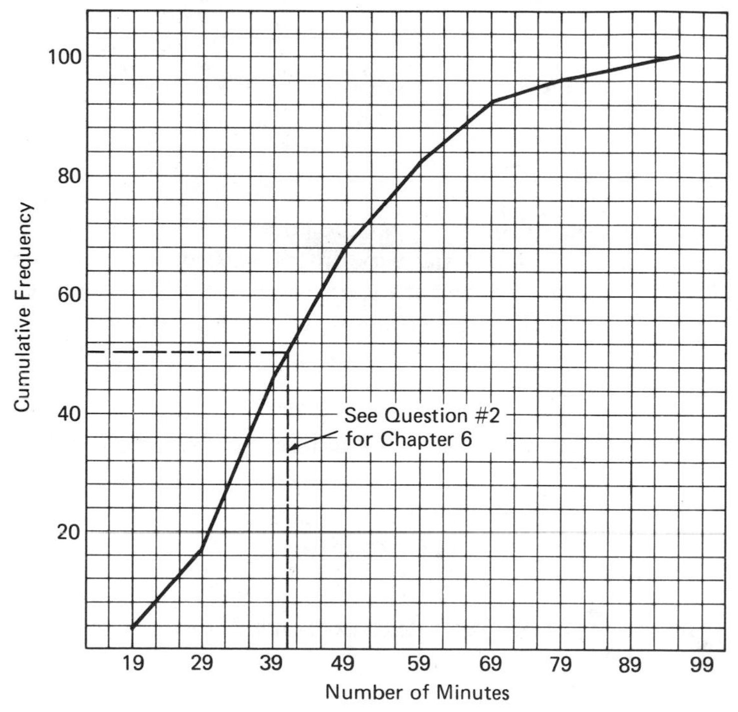

Figure N–8 Solution to Problem 5, Chapter 3.

CHAPTER 4. THE TASKS OF SCIENCE

Problems

1. This chapter devoted most of its attention to three types of experimental design, which I have called bivalent, multivalent, and factorial. One way to get a feeling for the meanings of these designs is to consider the relationship among them. If you can see them as forming a sort of hierarchy, it may help. By that I mean that a factorial experiment can be broken down into a set of multivalent experiments, which, in turn, may be broken down into a set of bivalent experiments. Imagine a factorial experiment with 25 groups made up of all combinations of five values of each of two independent variables. The table presents the design.

Values of Independent Variable B	Values of Independent Variable A				
	1	2	3	4	5
1	a	b	c	d	e
2	f	g	h	i	j
3	k	l	m	n	o
4	p	q	r	s	t
5	u	v	w	x	y

Continuing until you are sure you get the point:

a. Locate additional factorial experiments nested in this design.
b. Identify several multivalent experiments, of two different types.
c. Find several bivalent experiments of two different types.

Answers to Problems

1. a. A few additional factorial experiments, identified by letters in cells:

a b f g
f g h, k l m, p q r
p q r s t, u v w x y

b. The two types of multivalent experiment would involve just variable A; for example:

f g h
k l m n
u v w x y

or just variable B; for example:

a f k p u
i n s x
h m r

c. The bivalent experiments would involve just two values of variable A or two values of variable B.

CHAPTER 5. THE LAWS OF CHANCE

Problems

The following exercises will give you a little practice in figuring out the probabilities of binomial events. A good way to begin would be to construct your own version of Pascal's triangle. Use the spaces below

n		Number of Combinations
____	_ _	______
____	_ _ _	______
____	_ _ _ _	______
____	_ _ _ _ _	______
____	_ _ _ _ _ _	______
____	_ _ _ _ _ _ _	______
____	_ _ _ _ _ _ _ _	______
____	_ _ _ _ _ _ _ _ _	______
____	_ _ _ _ _ _ _ _ _ _	______
____	_ _ _ _ _ _ _ _ _ _ _	______

Figure N–9

(Figure N–9) and fill in the entries in the right- and left-hand columns, as well as in the main body of the triangle.

Before proceeding, you should check your triangle against Figure 5–3. Make corrections as necessary. Now, you can figure the probabilities asked for in the next sections. The answers will always take the following form. The probability of such and such an outcome is *a* in *b*, where *a* is the number of ways the outcome can occur and *b* is the total number of possible outcomes. I will set up all the problems in terms of heads and tails. A problem might look like this: The probability of throwing 5 heads in a single toss of 5 coins is *a* in *b*. The answer is $a=1$, $b=32$.

The problems to follow will be more complex because I will be asking about joint probabilities, which will require the use of the *either–or rule* and the *and rule*.

Either–Or Problems

1. The probability of throwing 8 or 9 heads in a toss of 9 coins is *a* in *b*.

2. The probability of throwing 8 or 9 heads in a toss of 10 coins is *a* in *b*.

3. The probability of throwing 2 heads and 1 tail or 2 tails and 1 head in a toss of 3 coins is *a* in *b*.

4. The probability of throwing all heads or all tails in a toss of 7 coins is *a* in *b*.

5. The probability of throwing at least 1 head in a toss of 6 coins is *a* in *b*.

And Problems

1. The probability of throwing 2 heads in a row is *a* in *b*.

2. The probability of throwing at least 3 heads in one toss of 4 coins and at least 3 tails in the next is *a* in *b*.

3. The probabilities of throwing 1, 2, 3, 4, etc., heads in a row are *a* in *b*, *a* in *b*, *a* in *b*, and *a* in *b*, etc.

4. In tosses of 10 coins, the probability of throwing an odd number of coins on the first toss and an even number of coins on the second is *a* in *b*.

5. If one person tosses 4 coins and another tosses 2 coins, the probability that the first will throw 4 heads and that the second will throw 2 tails is *a* in *b*.

A Note on Stud Poker

In the discussion of poker odds in Chapter 5, I mentioned that the order in which cards are dealt makes a difference in stud poker but not in draw poker. This is because of the way stud poker is played.

In the standard game of stud poker, each player receives five cards, one at a time, and there are rounds of betting after the second card and each succeeding card. The dealer deals each player a first card face down (closed card or hole card) and a second card face up. All the subsequent cards are dealt face up.

The right way to play stud poker involves one simple rule. If you are beaten on the table (by visible cards held by other players), get out. This rule relaxes somewhat as the hand proceeds. On the first bet, it is very strict: If your *hole card* is beaten on the table, and anyone bets, get out. Among other things, that means that you will sometimes drop out of a hand with an ace showing but a low card in the hole. The reason for this is that even if you pair the ace, you won't win much with such an obvious hand—and there is no guarantee that you will pair the ace.

Playing by this rule I usually win at this game. I recall two times when I have lost substantial amounts on a single hand, however, both times in the same way. I was dealt an ace down and another high card up and in order another of the high cards and two aces, giving me aces full, for example, ace, ace, ace, king, king. The winning hand in both of these disasters was the lowest straight flush, ace, deuce, three, four, five, of diamonds in one case.

The winner had been dealt the ace down and the other cards in the straight flush in an irregular order. By the time I had him beat on the table with a high pair, he had three in the same suit and still the possibility of pairing the ace. So relaxing the strict rule just a little, the winner stayed in for "just one more card," and eventually for the whole hand.

Obviously this player would not have stayed, even for a third card, if the first two had been low cards. This is an example of the point made in the chapter that order counts in stud poker.

Solutions to Problems

Either–Or Problems

1. $a=10$; $b=512$
2. $a=55$; $b=1,024$
3. $a=6$; $b=8$
4. $a=2$; $b=128$
5. $a=63$; $b=64$

And Problems

1. $a=1$; $b=4$
2. $a=25$; $b=256$ $5/16 \times 5/16 = 25/256$
3. $a=1$, $b=2$; $a=1$, $b=4$; $a=1$, $b=8$; $a=1$, $b=16$
4. $a=1$; $b=4$ $512/1{,}024 = 1/2 \times 1/2 = 1/4$
5. $a=1$; $b=64$ (16×4)

CHAPTER 6. THE NORMAL CURVE

Problems

The first of these involve additional analyses that you might do on the data presented in the end notes for Chapter 3.

1. Compute the average number of minutes seats were occupied in the cafeteria.

2. Using the cumulative frequency distribution for these data, use the graphic method (p. 225) to estimate the median.

The next problems involve two sets of numbers that happen to have a very handy set of features.

Set One	d	d^2
1		
2		
3		
4		
5		
6		
7		

Set Two	d	d^2
46		
47		
48		
49		
50		
51		
52		

3. Compute the means of these two sets of numbers.

4. Compute the standard deviations of these numbers, using the formula

$$S = \sqrt{\frac{\Sigma d^2}{N}}$$

5. What do these calculations tell you about the effects of adding a constant (45) to each of a set of numbers?

6. Compute Z scores for scores of 1, 4, and 7 in the first list and for scores of 47, 49, and 51 in the second.

The remaining problems involve areas under the normal curve. For later purpose it would be worthwhile to commit the following relationships to memory.

Z-Score Range	Percentage of Area
−.6745 to +.6745	50
−1.00 to +1.00	68
−2.00 to +2.00	95
−2.58 to +2.58	99

In order to deal with the following problems, you need to keep one more thing in mind: *The normal distribution is symmetrical about the mean.* The problems ask you to identify percentage of area within certain ranges for a normal distribution with these characteristics.

$$M = 50$$
$$S = 10$$

7. First, to review the relationship above, what limits would contain 50% of the cases in this distribution? 68%, 95%, 99%.

8. What is the limit above which 75% of the cases would fall? 25% of the cases?

9. What percentage of the cases would fall between 40 and 70?

10. What percentage of the cases would fall in the two tails of the normal curve above 70 and below 30? Above 75.8 and below 24.2?

11. Do you recognize the relationship of the answers above the levels of confidence?

Solutions to Problems

1. $M = \frac{\Sigma X}{N} = \frac{4{,}438}{100} = 44.38.$

2. See the cumulative frequency distribution in the answers to the problems in Chapter 3. The median is about 1/5 of the way into the interval between 39 and 49, something like 41. A check of the tally plot, counting cases from the low scores, confirms this. Note that the mean is higher than the median calculated above. This happens with positively skewed distributions.

3. The calculations look like this:

	X_1	d	d^2	X_2	d	d^2
	1	−3	9	46	−3	9
	2	−2	4	47	−2	4
	3	−1	1	48	−1	1
	4	0	0	49	0	0
	5	+1	1	50	+1	1
	6	+2	4	51	+2	4
	7	+3	9	52	+3	9
Σ	28	0	28	343	0	28
M	4	0	4	49	0	4

The mean of the first set of numbers is 4.0, of the second, 49.0.

4. $S=\sqrt{\Sigma d^2/N}$. The expression $\Sigma d^2/N$ (which you will encounter later as variance, S^2) is calculated as 4.0 for both distributions. Thus $S=\sqrt{4}=2.0$.

5. You will note that this value is the same for both sets of numbers. Adding the same constant to all numbers in a set of numbers does not change the standard deviation.

6. $Z_1=1-4/2=-3/2=-1.5$
$Z_4=4-4/2=0/2=0$
$Z_7=7-4/2=+3/2=+1.5$
$Z_{47}=47-49/2=-2/2=-1.0$
$Z_{49}=49-49/2=0/2=0$
$Z_{51}=51-49/2=+2/2=+1.0$

7. The first area is in the range .6745 standard deviation above or below the mean: $50\pm.6745\times10=50\pm6.745=43.255$ to 56.745. The others are 50 ± 1 standard deviation$=40-60$; 50 ± 2 standard deviations$=30-70$; 50 ± 2.58 standard deviations$=24.20-75.80$.

8. From question 7 you should see that 75% of the cases fall above 43.255; 25% fall above 56.745.

9. This is the area from 1 standard deviation below the mean (half of 68%, since the normal distribution is symmetrical) to 2 standard deviations above the mean (half of 95%). Thus 34%+47.5%=81.5%.

10. Since 70 and 30 are 2 standard deviations above and below the mean, and since 95% of the cases fall within these limits, 5% (the answer) fall outside them. For similar reasons the answer to the second part of the question is 1%.

11. These figures are exactly those involved in rejecting the null hypothesis at the 5% and 1% levels of confidence.

CHAPTER 7. SAMPLING THE UNIVERSE

Problems

One important set of ideas in Chapter 7 relates to sampling distributions and the effect upon such distributions of such variables as sample size. The accompanying table presents some data that will help to clarify these matters. The basic data, themselves, are the means of two sets of samples drawn from a table of random numbers. For one set $N=5$; for the other, $N=10$.

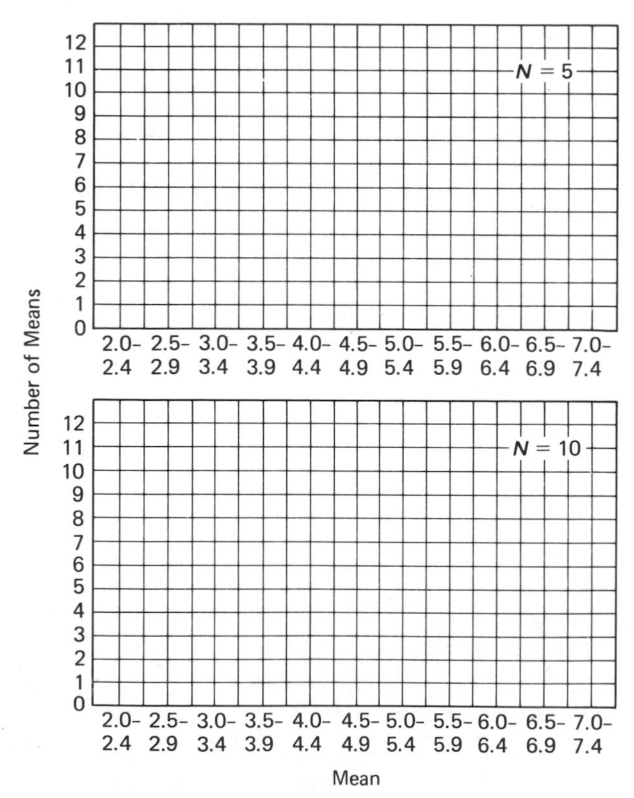

Figure N–10 Graph for Problem 1, Chapter 7.

1. On the coordinates provided (Figure N–10), make frequency distributions for these means.

a. What do these distributions represent?

b. Summarize the effects of sample size ($N=5$ vs. $N=10$) on these distributions.

Sample	Sample Mean ($N=5$)	Total Mean	Sample Mean ($X=10$)	Total Mean
1	3.2	3.2	3.4	3.4
2	4.2	3.7	5.5	4.4
3	3.2	3.5	4.0	4.3
4	5.6	4.0	4.3	4.3
5	3.6	4.0	4.2	4.3
6	4.6	4.1	3.2	4.1
7	2.0	3.8	4.2	4.1
8	6.0	4.0	3.5	4.0
9	4.8	4.1	3.8	4.0
10	6.4	4.4	6.3	4.2
11	4.8	4.4	4.6	4.3
12	3.0	4.3	4.9	4.3
13	4.4	4.2	5.2	4.4
14	7.0	4.5	4.4	4.4
15	7.2	4.7	6.2	4.5
16	4.8	4.7	4.5	4.5
17	3.2	4.6	3.7	4.5
18	3.8	4.5	4.1	4.4
19	6.2	4.6	4.5	4.4
20	5.8	4.7	4.3	4.4
21	6.4	4.8	5.9	4.5
22	4.0	4.7	5.1	4.6
23	3.4	4.7	5.5	4.6
24	2.6	4.6	2.9	4.5
25	4.2	4.6	4.0	4.5
Mean	4.58		4.49	
S_M	1.40		.88	
Expected S_M	1.28		.91	

2. The columns headed "Total mean" are the successive means obtained by thinking of the two sets of 25 samples as single samples that increase in size from 5 to 10, to 15, to 20, etc., in one case and from 10 to 20, to 30, to 40, etc., in the second.

a. What does an inspection of these means tell you about the accuracy of the estimated population mean as N increases?

b. What does this have to do with the law of large numbers?

3. Since we know that σ for a population of random numbers is 2.87, we can compute S_M for samples of 5 and 10 by the formula

$$S_M = \frac{2.87}{\sqrt{N}}$$

For the samples where $N = S$, we get

$$S_M = \frac{2.87}{\sqrt{5}} = \frac{2.87}{2.24} = 1.28$$

For the samples where $N = 10$, we get

$$S_M = \frac{2.87}{\sqrt{10}} = \frac{2.87}{3.16} = .91$$

The calculated estimates of 1.40 and .88 appear at the bottom of the table.

a. What does this tell you about the effect of sample size on the estimate of S_M?

b. What does this have to do with the assumptions underlying tests of statistical hypotheses?

4. On a completely different subject, what does this information about the effect of N on estimates of the mean suggest as a possible error in the reported value of a suburban home as $75,000, mentioned in the problems in Chapter 1?

Solutions to Problems

1. a. Sampling distributions of the mean.

b. The larger samples ($N = 10$) yield means whose distribution is more nearly normal. See Figure N − 11 for a couple of representations.

2. a. As sample size increases, the estimate of the mean becomes more precise. Where $N = 10$, the estimated mean is always within .1 of the known value of 4.5 after the 12th sample, a total of 120 cases. The data for samples of 5 are within .1 of the correct value after the 23rd sample, a total of 115 cases.

b. The law of large numbers was introduced in Chapter 5 in connection with proportions. As the number of cases increases, the proportion obtained comes closer and closer to the expected proportion. Obviously something very similar happens with statistics (e.g., the mean) of continuous measures.

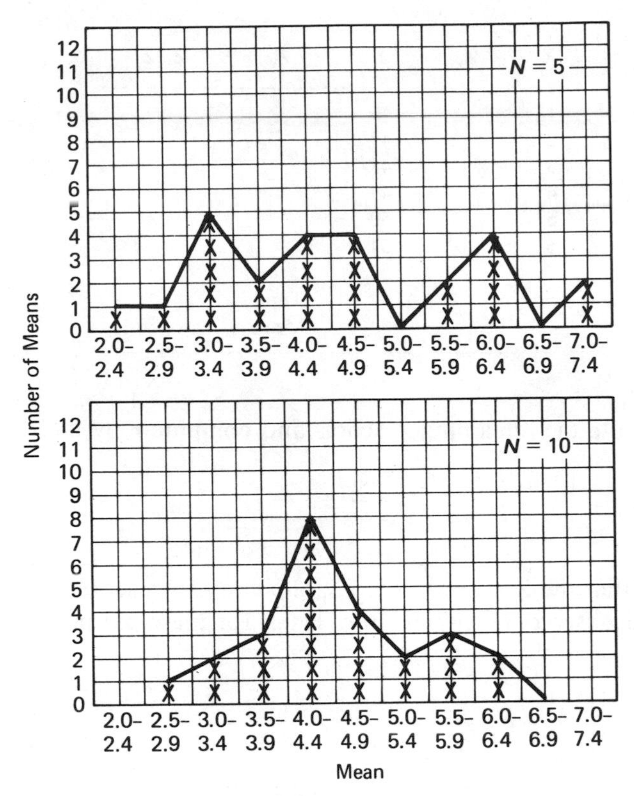

Figure N–11 Solution to Problem 1, Chapter 7.

3. a. Larger samples lead to a more precise estimate of S_M.

b. Small samples produce an inaccurate estimate of S_M. In this case the problem is complicated further by the fact that the sampling distribution of means is not normal and normal-curve statistics do not apply.

4. As a matter of fact, it was based upon just two houses. Obviously, such a mean is not very dependable.

CHAPTER 8. CORRELATION

Problems

In the problems for Chapter 6, you have already computed S and some Z scores for the following sets of numbers.

X	Z_x	Y	Z_y	$Z_x \cdot Z_y$
1		46		
2		47		
3		48		
4		49		
5		50		
6		51		
7		52		
			Total	
			Mean	

1. Compute the remaining Z scores and compute r by the formula

$$r = \frac{\Sigma(Z_x \cdot Z_y)}{N}$$

2. Now that the Z scores are available, you can rearrange the values of X or Y and observe the effects on r. Try it for these pairings:

X	1	2	3	4	5	6	7
Y	52	51	50	49	48	47	46

X	1	2	3	4	5	6	7
Y	48	50	52	46	51	49	47

The next problems involve two groups X and Y for which you have the following statistics:

$$M_x = 50 \qquad M_y = 100$$
$$S_x = 10 \qquad S_y = 15$$
$$r_{xy} = +0.60$$

3. What proportion of the variance in Y is accounted for by the correlation with X?

4. If someone has a score of 70 on X, what do you predict this person's score in Y to be? Remember $Z_x = rZ_y$, but this is in Z-score terms.

5. What is the value of $S_{y \cdot x}$? Remember that this is the square root of variance unaccounted for.

6. Given the value of $S_{y \cdot x}$ obtained in question 5, what are the 50%, 68%, and 95% confidence limits associated with the score on Y, predicted in question 4.?

(*Hint:* The answer to question 4 is between 115 and 120. The answer to question 5 is between 10 and 15. If your answers were not within these limits, you should check the answers and correct your mistakes before proceeding.)

Solutions to Problems

1.

X	Z_x	Y	Z_y	$Z_x \cdot Z_y$
1	−1.5	46	−1.5	+2.25
2	−1.0	47	−1.0	+1.00
3	− .5	48	− .5	+ .25
4	0	49	0	.00
5	+ .5	50	+ .5	+ .25
6	+1.0	51	+1.0	+1.00
7	+1.5	52	+1.5	+2.25
			Total	+7.00
			Mean	+1.00

Since r is the mean Z-score product, $r = +1.00$.

2. For the first problem, $r = 1.00$. For the second, the calculations look like this:

X	Z_x	Y	Z_y	$Z_x \cdot Z_y$
1	−1.5	48	−.5	+.75
3	−.5	50	+.5	−.25
5	+.5	52	+1.5	+.75
7	+1.5	46	−1.5	−2.25
2	−1.0	51	+1.0	−1.00
4	.0	49	.0	.00
6	+1.0	47	−1.0	−1.00
			Total:	−3.00
			Mean = r:	−.43

3. $r = +.60$, $r^2 = .36$, the proportion of variance accounted for. You may want to note that the answer would be the same if r were $-.60$.

4. Here are the steps:

$$Z_x = \frac{70-50}{10} = \frac{20}{10} = 2.0$$

$$Z_y = .60 \times 2.0 = 1.2$$

$$Y = M_y + (S_y \cdot Z_y) = 100 + (15 \times 1.2) = 118$$

5. Again the steps:

Proportion variance accounted for $= .36$.
Proportion of variance unaccounted for $= .64$.

$$\sqrt{.64} = .80$$
$$.80 \times 15 = 12 = S_{y \cdot x}$$

6. These problems involve an application of the information you have on areas under the normal curve. Thus 50% of the area is in the range $M \pm .6745S$; 68% of the area is in the range $M \pm 1S$. 95% of the area is in the range $M \pm 2S$. Now if you understand that the predicted Y of 118 is the mean of a distribution whose standard deviation $S_{y \cdot x}$ is 12, we arrive at

$$50\% \text{ confidence limits} = 118 \pm .6745(12) = 110 - 126$$
$$68\% \text{ confidence limits} = 118 \pm 1.0(12) = 106 - 130$$
$$95\% \text{ confidence limits} = 118 \pm 2.0(12) = 94 - 142$$

CHAPTER 9. USES AND MISUSES OF CORRELATION

Problems

1. The original study of "subliminal advertising" carried subthreshold directives which told the audience to "eat popcorn" and "drink Coca-Cola." Reports, never in quantitative form, were that these messages increased consumption of popcorn and Coke. Later indications were that this is an illustration of the base-rate problem. Can you think of why that might be so?

2. A follow-up study of genius children once reported that in 20 years since an original testing, the average IQs of these children had declined from about 145 to 115. The authors blamed the decline on lack of environmental stimulation. An aternative possibility is that the decline is partly, perhaps wholly, a regression artifact.

- **a.** Recalling that IQs are normally distributed and that normally distributed populations are produced by many independent causes, can you relate to this fact the phenomenon of regression to the mean?
- **b.** In the example above, the two IQ tests were very different instruments. Suppose, however, that the mean of both is 100 and that the standard deviation of both is 15. From the data provided, what was the correlation between the two tests?

The data in the accompanying table are the approximate batting averages of 18 major league baseball players, after 45 games and at the end of the season (B. Efran and C. Morris, Stein's paradox in statistics, *Scientific American*, **236**, 1977, pp. 119–127). As you can see, the 45-game averages are not very good predictors of the season average. The correlation between them is only +.23. The data are plotted in Figure N–12. The regression line relating predicted season averages to 45-day averages also appears on the graph. The numbers near the points are the numbers assigned the players.

Player	45-day Average	Season Average	Predicted Average
1	400	340	277
2	380	290	275
3	360	280	272
4	330	220	268
5	315	275	266
6	315	220	266
7	285	215	261
8	270	210	260
9	240	270	255
10	240	230	255
11	225	265	253
12	225	260	253
13	225	305	253
14	225	265	253
15	220	225	252
16	200	280	250
17	180	320	248
18	160	200	245
Mean	266	259	259

3. Would you call the correlation coefficient of +.23 a reliability coefficient, a validity coefficient, either, or neither? Why?

4. Suppose that after 45 days the managers of these players' clubs had decided to send every one back to the minors who was batting below 266 (the average for the first 45 games) and to retain every one who had been batting above 266. If a successful decision is the retention of people with a potential season average above 259 (the season mean) and the sending down of everyone who would have batted below 259, for which players would the decision have been the correct one?

5. Divide the incorrect decisions into two groups: injustice to the players and injustice to the club.

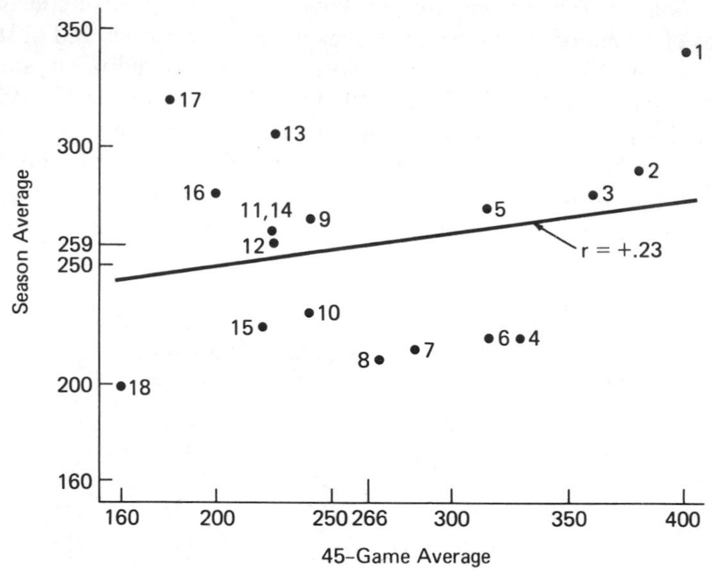

Figure N–12 Graph for Problems 3 through 6, Chapter 9.

6. Stein's "paradox" referred to in the title of the article from which these data came is that, in most statistical situations, past performance provides the best prediction of future performance, but in this case it does not. In this case, the prediction from the correlation coefficient is better for 14 of the 19 players.

a. Identify them.

b. What concept does this illustrate?

c. What would interpretations in terms of factors that cause players to go into and come out of slumps indicate?

d. If you are a gambler, you can win some bets by making predictions of future performance early in the season. How?

Solutions to Problems

1. What seems actually to have happened was that the temperature was lower than average on the experimental days. This led people to eat popcorn, the saltiness of which increased the consumption of Coke.

2. a. Extreme scores (high or low) result from chance combinations of extremely favorable or unfavorable values of the many independent causes. Later measures are not likely to take place under such conditions, and performance is less extreme.

b. Since the original IQ was 3 standard deviations above the mean and the second IQ only one, the formula $Z_y = r \cdot Z_x$ required that $r = +.33$; that is, $1.0 = .33 \times 3.0$. Although the tests involved in the study were very different, it is unlikely that the correlation between them was less than +.50 or so. Some part of the lowered IQ probably is a real effect.

3. Either, I suppose. It is a validity coefficient in that it predicts a performance of interest; but it is also a (low) reliability coefficient because it is based on two successive measures of the same thing. As with most sharp distinctions, this one dulls at times.

4. The decision would have been correct only for players 1, 2, 3, 5, 10, 15, and 18.

5. Injustice to the players: 9, 11, 12, 13, 14, 16, and 17.
Injustice to the club: 4, 6, 7, and 8.

6. a. The correlation predicts better for all players except numbers 1, 10, 15, and 18.

b. Regression to the mean, because the better prediction from r (very low) is no different from the accuracy you would get by predicting the mean for all players.

c. Regression error.

d. By betting that the leaders will lose points in their batting averages and that weak batters will improve.

CHAPTER 10. ANOVA

As a first project for this chapter, a review of the discussion of factorial designs in Chapter 4 might be in order. Having finished this one, you should be in a better position to appreciate the treatment presented there.

Problems

Suppose that someone does a study of crime rates in three different types of neighborhoods (A = poor, B = middle income, C = rich) provided with three different types of police activity (1 = same as before the experiment, 2 = more lenient than before the experiment, 3 = stricter than before the experiment). Suppose that there are five neighborhoods of each type—five observations per cell. Finally, suppose that the results are as follows, where the data are average crimes per thousand of population: A-1, 40; B-1, 35; C-1, 30. A-2, 35; B-2, 30; C-2, 25. A-3, 45; B-3, 45; C-3, 45.

1. Make a diagram for this study and the results.
2. Complete the following portion of an ANOVA table.

Source of Variation	d.f.
Total	

3. Suppose that all differences in the data, collapsed across conditions where necessary, are significant.
a. Is there a main effect of neighborhood?
b. Is there a main effect of policy action?
c. Is there an interaction?

Suppose that a consumer's organization makes a comparison of three foreign compact cars made in Italy, Japan, and Germany. They rate handling, quality of ride, effectiveness of brakes, and hill-climbing power on a scale where 4 is excellent, 3 good, 2 fair, and 1 poor. A single sample of each car (this gets to be important) is bought anonymously from local dealers. Suppose that the ratings are as follows.

	Aspects Rated				
Car	Handling	Hill Climbing	Brakes	Ride	Mean
German	3	3	4	3	3.25
Japanese	2	2	3	2	2.25
Italian	1	1	2	1	1.25
Mean	2.0	2.0	3.0	2.0	

4. a. Does this table show a main effect of aspect rated?
b. Is there a main effect of type of car?
c. Is there an interaction?

5. Fill in the ANOVA table below. You are probably in for a surprise here. Part of the right answer will seem wrong. It is not.

Source of Variation	d.f.
Within groups	
Total	

6. Suppose that you are primarily interested in the quality of ride in your selection of a compact car. What confidence can you place in the superior rating of the German import?

Solutions to Problems

1.

Type of Police Action	Type of Neighborhood			Total
	A	B	C	
1	40	35	30	105
2	35	30	25	90
3	45	45	45	135
Total	120	110	100	

2.

Source of Variation	d.f.
Neighborhood	2
Police action	2
Interaction	4
Within groups	36
Total	44

3. a. Yes
b. Yes
c. Yes

4. a. Yes
b. Yes
c. No

5. Since there is only one sample of each car, the within-groups d.f. are zero.

Source of Variation	d.f.
Aspect rated	3
Type of car	2
Interaction	6
Within groups	0
Total	11

6. Very little. Since there are zero degrees of freedom within groups, you cannot generalize any single rating to the population of cars. The fact that four aspects were rated, however, does make it possible to make the (possibly more important) generalization that, overall, the German car is superior. This example illustrates an important problem with evaluations of expensive products. If only one example is sampled, you cannot generalize to the population, unless you have reason to assume absolute standardization in the product. Such as assumption is dangerous. There really are lemons in manufactured products. In the case of automobiles, those turned out just before or just after a weekend have a higher probability of being defective.

Index

Entries following a notation, *d.* refer you to a definition of the term in a glossary.

T

U

V

W

Z